WALTHER CLOOS/THE LIVING EARTH

WALTHER CLOOS

THE LIVING EARTH

The Organic Origin of Rocks and Minerals

Translated by K. Castelliz and
B. Saunders-Davies
from *Lebensstufen der Erde*

Lanthorne Press
1977

Translation of the second enlarged edition in German of *Lebensstufen der Erde* by W. Cloos, Stuttgart 1970

Published by permission of Verlag Freies Geistesleben, Stuttgart.

ISBN 0 906155 01 0

Printed and bound in Great Britain by
R. J. Acford Ltd., Industrial Estate, Chichester, Sussex

Contents

List of Illustrations

Illustrations, except where otherwise stated, are by the author.

Plates are between pages 80 and 81.

Translators' Foreword

The increasing number of enthusiasts who are attracted by the rich variety of rocks and minerals around us may find this book difficult to lay aside. Walther Cloos' train of thought demands an almost meditative approach; for the mind has to live in pictures of life-processes very different from those of the present day. But how the old rocks come to life as one looks on them with new eyes!

'The Living Earth' is at present the only book of its kind available to the English-speaking reader. As a companion work to Grohmann's 'The Plant' it should be helpful to teachers in Waldorf Schools and to those concerned with the soil and Bio-Dynamic agriculture.

The translators wish to express their thanks to the Verlag Freies Geistesleben for the use of the blocks for the illustrations, and to Selma Hedges for her care in producing the admirable cover design. They also warmly thank John Docherty for reading the text and for his help with many geological details and the table of formations.

Author's Preface

In this book we have attempted to look at scientific facts concerning the world of rocks in the light which Rudolf Steiner has shed on all spheres of human life and endeavour in his anthroposophy or spiritual science. Seen from this point of view the Earth is a living organism whose life is interwoven spiritually with the life of the heavenly bodies in the solar system and the whole realm of the stars. What has developed gradually within this life as the kingdoms of nature – mineral, plant and animal – are stages left behind by the being which is the culminating goal of all Earth existence – MAN.

The author is aware that his attempt to arrive at a comprehensible picture of the world of rocks contradicts the view currently held. The fact that this view has lead to a 'Rape of the Earth' and the threat of atomic death may excuse the attempt with all its shortcomings. If we would prevent the premature end of the Earth through arbitrary and thoughtless actions of the present-day, we must come to a new conception of man and the Earth.

We would refer the reader to the works of Günther Wachsmuth especially his book 'Die Entwicklung der Erde' which deals with the time element of our subject matter.

I would like to thank all those friends and colleagues who, through discussions over many years, have contributed consciously or unconsciously to bringing this book to birth.

WALTHER CLOOS

Stuttgart, Midsummer 1958.

Introduction

The wayfarer who turns his gaze from the abundant life of plants and animals to the solid mineral earth, may not always realise that traces and remains of bygone life are to be found in the rocks and screes. Only in comparatively few places on the surface of the Earth can one see that the rock contains fossilised remains of animals resembling mussels and snails. These and related forms constitute by far the greatest part of what is clearly recognisable as traces of life. Furthermore it is evident that the rocks containing such traces of life are mostly limestones or calcareous shales or marls, or shales inter-bedded in limestones.

If one takes the trouble to examine limestone under a magnifying glass or a microscope, it is easy to see that many consist entirely of the minute tests of protozoa (protista), organisms which inhabit our oceans to this day. The zoologist Hertwig writes: 'No group of animals has contributed so much to the formation of new strata in the present and in the past as they have.'

We find the protozoa in layers between the coal seams as fusulinid limestone and we find them above all in the Muschelkalk, in the Jurassic limestones, the Chalk and the nummulitic limestones of the early Tertiary period.

If we think of the plateau mountains of the Jura, the white chalk cliffs of the south coast of England and the Baltic coast near Rügen and further afield the endless central plains of the American continent which are underlaid by thousands of metres of chalk, we have to realise that these tremendous layers of limestone are largely made up of minute organisms which lived in myriads and have perished.

Certainly, we still find today in the oceans, myriads of these little creatures which deposit their calcareous shells (tests) to form the ooze of the deep sea bed, but we can no longer observe the formation of tremendous calcium deposits like those of the past. It is obvious that the abundance of life is reduced, that the intensity of these processes has slackened off and also that these little creatures tend to be much smaller than they were in the past.

All these observations suggest that, at the time of the formation of these limestones, life processes were far more all-embracing, that the oceans, where they occurred, predominated and that the Earth as a whole was in a more watery state. This idea is substantiated by the fact that in some of these limestone strata we find the remains of large animals which at first somewhat resemble fish, newts, crocodiles, and later lead on to the strange forms of the giant saurians.

From the change in these forms of animal life we can see that this time was a significant transition period. The dominion of the sea was broken by the gradual formation of solid land. As we have seen, it was the tiniest organisms which were mainly responsible – at least in these limestone strata.

If we turn from the Chalk, Jurassic limestones and Muschelkalk, and go deeper into the Earth, we observe that the limestone layers become scarcer and we find rocks of which the blue roofing slates are a typical example. In these slates we find remains of marine animals, many of which today are extinct. From these animal forms it is clear that, though water played an important role when the rocks were formed, it must have been a very different sort of ocean, for it did not only leave limestone beds but also clayey mud, for slate consists mainly of aluminium silicate with the admixture of carbon and iron. The slates alternate with sandstones, or sometimes with limestones or calcareous shales. As one penetrates deeper, approaching the earliest times of the Earth, their place is taken by the glittering mica-schists and gneiss. At the end of the schist/slate period something very special appears – the Coal Measures. They indicate a more vegetative life in contrast to the animal life of the limestone period.

If one goes still deeper and further back beyond the schist and gneiss periods, one comes to the real primal rock exemplified in granite and its related rocks. These rocks are granular and more or less uniformly composed of quartz, mica, hornblende and felspar. It is important to bear in mind that the oldest rocks like granite have a granular structure, the schists and slates and shales are foliated and limestone is characterised by thick beds.

In the deeper layers of shale, embedded in between limestones, sandstones and salt deposits lies one of the most important remains of ancient life – oil. This however constitutes only a part of the oil. Oil reappears again in the Tertiary epoch. It is the same with coal; it reappears in the Tertiary as brown coal.

In the depths where the rocks begin to become crystalline (mica-schists and gneiss, Plate 1) the last traces of life in the form of fossils and impressions disappear. But even here we find between the strata huge beds of marble or dolomite (a double carbonate of calcium and magnesium) which point towards tremendous plant and animal life processes and may have been produced by something similar to our present day algae (seaweeds).

With this rough sketch of the three principal periods of the Earth on which our further studies will be based, we have by no means covered all the rocks that constitute the hard crust of our Earth. To these must be added the group known as igneous.

These igneous rocks of which we will only mention granite, the greenstones, porphyry, trachyte, plateau basalt and lava (Plates 2, 13, 14) have from the most ancient times domed up, broken through and overflowed the layers of sandstones, slates and limestones. They are closely associated with the formation of mountain ranges, though of course not all mountains are of volcanic origin.

To get an idea of the mass of the various strata we need a few figures. The American F. W. Clarke has determined this for the known part of the Earth's crust to a depth of 16,000 metres. This depth can be considered as known since oil borings have reached 6,000 metres and more in many parts of the world, and tremendous folding has brought the deep rocks to the surface in mountain ranges many thousands of metres above sea level. If one takes into account the mass of the rocks, of the sea water and the atmosphere, one finds that the comparative weights are: Earth's crust 93.06%, water 6.91% and atmosphere 0.03%.

Of this 93.06% of the Earth's crust, about 88.4% consists of igneous rocks, 3.7% of schists, slates and shales, 0.7% of sandstones and only 0.2% of calcareous rocks. It is evident that the rocks we have classified as of granular structure, far outweigh everything else.

The relationship between the igneous rocks and the sedimentary rocks (sandstones, limestones, slates, etc.) appears to be very different if one only considers the surface of the Earth, as some three quarters of it is covered with sedimentary rocks and only one quarter has the igneous rocks exposed.

Again if we consider the thickness of the individual strata their massiveness is such that one simply asks: Where has it all come from? If one looks at all these phenomena with an open mind one is struck with amazement and wonder, and it is this attitude of mind which *should* be the starting point for true scientific enquiry.

*

The question of the origin of the rocks is as old as science itself. The age old dispute between the neptunists who maintained that everything arose from water, and the plutonists who referred everything back to primal fiery and molten processes, is not settled even today. The modern assumption is basically that rock materials originated from a condition of incandescent gas. The gas would then have cooled to a molten mass producing a 'primal magma' – somewhat resembling basalt. After further cooling, water would have formed on the surface crust, and from its interaction with the boiling magma of the interior, the enormous variety of rocks arose.

This theory considers the sedimentary rocks of the slate/limestone times as simply made up of debris and dissolved matter ultimately derived from the cooled magma. Even the enormous masses of calcium used by the protista to build their tests are explained as originating from the calcium content of the primal magma dissolved in the sea water.

This theory of the origin of rocks leaves a cardinal question unanswered: *Where did life come from?* The origin of life is a question to which modern science has absolutely no satisfactory answer. There was, and still is, a theory about bacteria in the universe which, through pressure of sunlight, entered the cooled atmosphere and so formed the first life on Earth. More modern theories postulate a spontaneous generation of simple proteins from methane, carbon dioxide, ammonia and water in the warm atmosphere through electrical discharge. From these bacteria-like or virus-like primal proteins began a development which, assisted by influences from outside, finally produced man.

There are several theories of spontaneous generation but they all forget the most important thing. *There is not only life on Earth but even ensouled life, moreover ensouled life that is aware of itself as a spiritual being – Man.*

If one considers man on earth endowed with soul and spirit, and tries to reconcile this with scientific views of geology, one comes to an impasse. In practice this is so because today we normally only observe physical and chemical processes at work in the rocks, and no longer life processes – albeit the calcareous mountains bear witness to former life processes.

In plant, animal and man the mineral is a living, feeling and thought-bearing substance and as such it is not subject to the chemical and physical laws that hold sway in the world of rocks. But when a plant, animal or man dies and lays aside its body, then the chemical and physical laws take over and dissolve it.

If there were a tiny creature able to study the decaying processes of a corpse, yet unable to see the whole and understand the organic origin of this corpse, it would be in the same situation as present-day man vis-a-vis the rocks of the Earth. In the same way as this little creature would only find physical and chemical laws active in the decaying corpse, so in the world of the rocks today we can only observe physical and chemical processes. The question of the origin of the mineral mass would preoccupy it when it came across the quantities of calcium phosphate which constitute our skeleton. It would also find in the corpse iron, magnesium, sodium and carbon – all materials which are widely distributed in our rocks. If however this little creature observed that all these substances were arranged in a certain *order,* that there were layers and zones in which one or another substance predominated or receded, it might arrive at the conclusion that behind this order there was an 'idea', a principle which had ordered it.

Facing the world of rocks we find an ordering. Goethe in his geological studies came to the conclusion that the perfect classification is not made by man; Nature herself has laid it out before his eyes. This classification or order – obviously recognised by modern geology – has to be further pursued without letting one's gaze be dimmed by preconceived ideas and trying *to explain the past from the present.*

Rudolf Steiner in his cosmology has so presented the 'idea' behind the formation of the world of rock, that one can see how *the mineral deposits of the Earth are the transformed remains of former life.*

This former life, which had no resemblance to that of today, has transformed itself into the present higher kingdoms of nature but it has also left its 'corpse' behind. If we would not disect this 'corpse' but look at it as a whole, then the original magnitude of this ancient life would emerge and the layers of the past would become the leaves of a book which we could learn slowly but surely to read.

1 *The Basic Framework of the World of Rocks*

Wherever the oldest crystalline rocks of the Earth are found, they show very little variation the world over. Be they on the surface or deep within the Earth, they are simple combinations of quartz, felspar and mica or hornblende. The differences are mainly in the size of the crystalline granules and the colour. The colour depends on the relative amount of mica and felspar, and on certain metals, mostly iron and sometimes manganese.

This uniformity is very characteristic of these oldest times. Differentiation only starts later. Thus we have a time in the history of the Earth when neither slate nor limestone nor coal were formed. There was only one 'basic process' which we can call *granite formation*.

Immediately above the granite and also inter penetrating and surrounding it one finds the first modification of the crystalline structure in gneiss. Gneiss is not so uniformly structured as granite. Though still granular, it is layered. It can be relatively homogenous with layers caused by the little flakes of mica, or it can consist of layers of quartz, mica and felspar more or less separated from each other. One has the distinct impression that what has developed as mica has given the whole rock its structure. This 'mica-effect' is the hallmark of the development of the following period just as the granular structure was for granite.

A further development of this separation is to be found in mica-schist where the felspar has almost completely disappeared and the quartz lies in thick layers in the schists. At the same time a further development of granite itself (quartz, felspar and mica or hornblende) takes place. These new granitic rocks, the greenstones – syenite, diorite, diabase, gabbro and serpentine – contain no quartz. Syenite and diorite are still very granitic so that the layman can easily confuse them. All these rocks consist of felspar and mica or hornblende, and can occur between gneiss and mica-schists. Magnesium plays an important role. The dark iron-containing mica of syenite, and the green hornblendes and related minerals are primarily magnesium ferro-silicates. These latter give the rocks, especially diorite, diabase and gabbro, their green or grey-green colour.

There appears to have been a conflict at this stage between two formative tendencies – one to form laminated schistose rocks, the other granitic granulated ones. So it happens that syenite, diorite, diabase and gabbro can appear laminated or granular.

Thus it is incorrect to speak of granite as an igneous rock as used to be done, for what we call today the 'igneous process' has itself gone through a *development* like everything else in nature. The beginning of this development is to be found in gneiss, in the crystalline schists and the greenstones. Later we shall see that it is from these rocks, especially the greenstones, that the true igneous rocks such as the basalts and other lavas develop. As the third stage of this development, which one might consider as a differentiation of tendencies within the primal granite, we have the appearance of sandstones, slates and limestones. Granite is composed of quartz, felspar, mica and/or hornblende. What was quartz now is sandstone, what was mica or hornblende appears as slates and shales, and in the place of felspar we have limestone. (Mica contains much aluminium and magnesium; and in granite, felspar is the bearer of 'limestone forces'). The threefold composition of granite becomes the great threefold division of sedimentary rocks – sandstones, shales and limestones.

The granite appears as a sort of Proteus, a forerunner containing the possibilities of future development. The processes of rock formation become more and more diversified and as the primaeval unity disappears, the primal granular rocks also go through a further development. We have outlined how, as the quartz more or less disappeared, granite became syenite, diorite and all the other granular greenstones.

After this 'greenstone stage' quite new forms of rock appear – the porphyries (Plate 2). These porphyries can sometimes look like granite but are generally more finely granulated or even glassy, with the felspar in large crystals. They contain little or no mica, but hornblende and quartz instead, which in true porphyries are blended in a glassy mass.

These porphyries appeared about the same time as the first sandstones, shales and limestones. Later on limestones take precedence over sandstones and shales. We know that limestone is the result of animal life. It is thus a true sedimentary rock, as are also the accompanying sandstones and shales. At this stage the 'eruptive rocks' trachyte, basalt and lava also occur. Their origin will be discussed later.

*

We have endeavoured to reduce to its essence the great complexity of rocks. All rocks are distributed over the Earth in distinctive ways and there are large areas where one or other are entirely absent. Thus it is clear that the original uniform process that produced granite became differentiated, so that in one area the development was in the direction of laminated rocks (schists, etc.), in another towards limestones and in another towards the greenstone-type rocks.

We may tabulate this as follows:

1. **Granite:** granular.
2. Gneiss: beginning of laminated structures; **greenstones**: both granular and laminated.
3. Sandstones, **shales** and limestones: the granite formation has broken up.
   Slates/shales predominate.
   Porphyries.
4. Sandstones, shales and **limestones**: **limestone** predominates.
   Trachyte, basalt and lava.

Based on the composition of granite we have:

quartz	mica hornblende	felspar

In the sedimentary rock development this becomes:

sandstones	shales	limestone

The corresponding 'igneous' rocks according to the above presentation are:

(Older formations) granites	greenstones	porphyries
(Younger formations) trachyte	basalt	other lavas

Such an arrangement does not correspond with that of orthodox geology. It does not attempt to be complete, but it deals with the basic phenomena. Such a skeleton outline can be a great help to find one's way later through the complexity of the phenomena. If one only considers these simple phenomena, of course one comes up against a difficulty when one considers limestone and coal formations. We know that limestone is the product of animal life and that coal originates from plant life. Both deposits are to be found in old and young strata and always lie between apparently entirely inorganic layers. These inorganic layers of sandstone, shales or igneous rock appear as though they originated in the same way as what we observe today when sand or mud are deposited in water, or the molten lava flows from a volcanic crater. But today sand no longer makes sandstone and clay does not form shales, neither does lava turn into granular crystalline rock.

We have already seen that the enormous quantities of limestone can only be understood if we postulate a far more intense life then exists today in our seas.

Likewise with coal. It has been calculated that a tropical rain forest of today would, if suddenly destroyed by a natural catastrophy, only yield a layer of coal a few millimetres thick. Here we are faced with the question: What an extraordinary plant life it must have been to yield hundreds of metres of coal.

Today we know of no life process of such intensity that it could form hundreds of metres of limestone or coal within any conceivable time. Were there not perhaps quite different conditions then on the Earth which do not follow the natural processes of today? Modern research does not offer any satisfactory answers. The answers given speak only of immense periods of time, high pressure in the interior of the Earth, high temperatures which have transformed materials. Experiments appear to support these theories. But experiments alone do not satisfy, since they take as their point of departure lifeless substances which are taken to be the origin of everything.

Nowhere can we observe that life develops from inorganic substances. On the contrary we see that life decomposes *into* inorganic substances which do not come back to life. Living organisms can take up inorganic substances but not all these substances need necessarily have been of mineral origin.

In modern biology the experiments of Warburg show that plants have the capacity to collect extremely finely dispersed substances in the atmosphere or soil (physics speaks of highly ionized conditions) and condense them into measurable quantities. In the Bio-Dynamic agriculture introduced by Rudolf Steiner, this fact has been made use of for more than 40 years. This method shows quite clearly that the living soil with its plants has the capacity to replenish its basic nutrients without the application of mineral fertilisers. If one tests soils treated Bio-Dynamically for 40 years one finds that phosphorus, potassium and nitrogen are present in the necessary quantities in both soil and plant. Although the plants appear to take up these substances year after year from the soil, yet they do not diminish.

The trouble is that through the use of mineral fertilisers the plant can get out of tune with the original order of life. A plant which, for decades and many generations, has been bred on exhausted mineral fertilised soil has simply lost the capacity to collect its own substances; it has degenerated, and follows those laws which the experimentor has imposed upon it. This is the secret of one of the fundamental errors of our time, and this error casts its shadow even into our ideas of the past evolution of our Earth.

If these ideas lead to the conviction that *first* there was Life and from it condensed the inorganic and lifeless, then we are in a position to take the researches of Rudolf Steiner on the past of the Earth as the foundation of our further observations.

2 *Previous Stages of the Living Earth*

If one presumes – according to Rudolf Steiner's spiritual investigation – that first there was life, and the dead mineral matter was cast off as part of a dying process, the question arises: From where comes life? The answer to this question does not come within the scope of our discussion for we are dealing especially with the origin of the mineral kingdom and its forms. For the answer to this question we would refer to Rudolf Steiner's cosmology in his book *An Outline of Occult Science*. The past stages of the Earth which Rudolf Steiner there describes must however concern us, for they are the basis for those vital earth-processes that lead the not-yet-tangible into the mineral, visible form. These processes also underlie Ernst Haeckel's biogenetic basic law which states that the development of individual organisms repeats the development of the world as a whole. (As formulated by Haeckel the development of the embryo is the repetition of the development of the whole tree of life.)

The prehistory of the Earth is repeated in the embryonic development of plant, animal and man; this alone allows them to appear in their present form. This recapitulation is obscure and not directly observable, for it takes place in the smallest and most delicate realms of life. Rudolf Steiner suggested that through a study of the physiological processes of embryonic development it would be possible to gain a true picture of the whole past development of the Earth.

It will be the task of the future to discover the secrets of this past Earth development especially through the study of human physiology.

With regard to the mineral kingdom it is different. Since its forms are really cast-off remains of life we are dealing with something that is finished and not with something in the *process of becoming,* the physiological processes of which we could investigate. We may however be sure that in this 'wrought work' which has come to a certain state of rest as mineral kingdom we will be able to find images which can tell us something of the life processes of the Earth's past as described by Rudolf Steiner.

What we find in the mineral kingdom shows that in the younger layers of sedimentary rocks remains of animal life-processes predominate, whilst in the older ones we have mainly remains of plant life-processes (coal). In the oldest rocks all traces of life are completely absent.

This only informs us that some of the living organisms were sufficiently dense to remain preserved. It does not mean that there were no other forms of life present. The apparent absence of traces of life in the primitive rocks is only an absence of those forms which we know as the manifestation of life in later times. How can one relate the life processes of the earliest beginnings of the Earth – before mineralisation – to what later is precipitated and becomes rock? We can gain some idea by observing how in animal and man one has the metabolic processes on the one hand, and on the other, the end product of these processes deposited as substances such as bones, nerves, brain, etc. These are, by comparison with organs such as the liver, lungs, kidneys and heart, endowed with only a minimum of life. For example portions of nerve and brain damaged through injury are not capable of regeneration.

In the plant world we have comparable processes in the flower and seed formation on the one hand, and wood and root formation on the other. Flower and seed formation correspond to metabolism in man; wood and root to bones, nerves and brain.

From such living examples we can see how life processes are related to structures which have been cast off from the living. In other words, one cannot adequately consider the mineral kingdom without taking into account the plant, animal and human kingdoms. These have, in the course of their development, deposited the mineral kingdom in the same way as man and animal form their bones, nerves and brain, and as trees form their wooden trunks.

Anyone who considers the mineral kingdom alone without including the life processes which have formed it, is like someone trying to explain bones, nerves and brain in isolation, or like a timber expert who does not ask how or where timber grows.

In 1923 Rudolf Steiner wrote in one of his note-books something which links the mineral world with the development of the kingdoms of nature and man. What the note contains is the result purely of spiritual scientific investigation and expresses in a few words the development of the whole of nature and man. Rudolf Steiner referred to this theme in numerous lectures in the preceeding decades, but nowhere is it dealt with thoroughly and in detail. One has the impression these notes are a sketch for a cycle of lectures which were never given. The text* reads as follows:-

In porphyry the world-plant-animal expires,
then in schist the plant-nature,
and in limestone the animal-nature.
In salt the universal human being is extinguished.
The opposite pole is sulphur in which the mineral is burnt and consumed.

Mineral is consumed in sulphur,
Plant spreads warmth in layers in schist,
Plant-animal in-grains feeling in porphyry,
Animal conserves forms in limestone,
Man condenses thoughts in salt.

With fiery sulphur man enters the earth,
In schistose layering he makes himself at home on earth,
By awakening feeling (porphyry-like) he finds his limbs and forms himself as man in limestone,
To found the ground for thought in salt-depositing.

Im Porphyr erstirbt das Welten-Pflanzen-Tier –
dann erstirbt im Schiefer das Pflanzenwesen
im Kalk das Tierwesen –
und im Salz erlischt das Menschenwesen –
Der andere Pol ist der Schwefel in dem das Mineral verbrennt.

Mineral verbrennt im Schwefel
Planze schichtet Wärme in Schiefer

*First published in 1927 in the weekly *'Anthroposophie.*Wochenschrift fur freies Geistesleben' 9th year. No. 41.

Pflanzen-Tier körnt Empfindung im Porphyr.
Tier bewahrt Formgebilde im Kalk
Mensch dichtet Gedanken in das Salz.

Sulphurisch tritt der Mensch ins Irdische
Schiefrig macht er sich die Anpassung ans Irdische zurecht
Empfindung weckend (porphyrartig) gliedert er sich und
formt sich menschlich im Kalk,
um in Salz-Einlagerung die Gedankengrundlagen zu schaffen.

Here a concrete connection is established between life processes and rock formation. New concepts emerge such as 'plant-animal' and burning up of minerals in 'sulphur'; and then there is the significant pointer as to how man is connected with the five main processes active in the formation of rocks. We shall go further into these new concepts later in order to understand what is at first sight new and rather startling.

The first part of the above text speaks of a dying away as the basis of the formation of rocks. One should not imagine that these 'world-plant-animals', 'animal-beings', 'plant-beings' and 'human beings' were single living individuals which died in masses and formed rock out of their bodies. When we speak in this sense of the 'world-plant-animal' and of the being of plant, animal and man, we are referring to something universal which contains these kingdoms of nature but as yet not differentiated individually. This 'being' is all-embracing in the sense that its life forms the whole Earth. The Earth itself was once living as a world-plant-animal, it was entirely plant being, animal being, human being. As this original all-embracing entity dies and disappears, there emerges that which later becomes rock and the *individual* forms of plants, animals and man.

In order to imagine spacially this primaeval condition of the Earth, one has to picture the Earth at that time as having a circumference extending far into the planetary spheres. In a lecture to the workmen at the Goetheanum of September 23rd 1922, Rudolf Steiner states that the Earth was enormous and 'in fact approximately as large as Jupiter'. The Earth has shrunk tremendously, it has dried up, withered and its former life is dead. Originally this life was active in the periphery. In the process of dying and shrinking, the residues of this life rained down from this periphery and later condensed and hardened as rock.

The beginning of the shrinking of the Earth goes back to a far distant past. The major stages of the Earth evolution described by Rudolf Steiner in his *Outline of Occult Science* as the Saturn, Sun and Moon conditions were extended spacially to the present-day orbits of Saturn, Jupiter and Mars. In this process of contraction our present planets were separated off and left behind, as was the Sun. The original body of warmth of Old Saturn contracted and became the gaseous body of the Old Sun and this contracted further to the fluid body of Old Moon. Before each new development, a recapitulation of previous stages of these past incarnations of the Earth took place according to the biogenetic law.

At the beginning of our present Earth development there was also a recapitulation of these previous stages. Then that which was peculiar to our Earth emerged, and this led finally to the formation of solid rock. Thus we see that we are dealing with rhythmic processes or repetitions each on a higher level than the last.

The first beginning of the Earth is a pure heat phenomenon. Warmth-substance, sacrificed by a hierarchy of creative beings, is the origin of the material development of the Old Saturn stage. The 'result' of this Saturn-development was the emergence of the germinal stage of man in the form of entities without individual life. This life was only incorporated on the Old Sun. At the same time heat densified and rarified. Gas and light emerged. On the subsequent Old Moon stage man was endowed with feeling, the gaseous was densified to the fluid state, and on the other side, sound came into being. Only during the Earth stage did man begin to develop a human individual consciousness, and the *solid* separated from the fluid. Further there emerged the form-principle, born of sound, capable of shaping the solid physical substance.

So we can look back on a sequence of evolutionary events whose nature and complexity unfold rhythmically, rising higher at each repetition. The original heat substance which really was the beginning of what later became the solid mineral, experienced during the second stage an infusing of life. The plant kingdom was conceived. During the third stage this substance was endowed with feeling and became the first beginning of the animal kingdom. Only during the fourth stage, on the Earth, did this substance become bearer of conscious thought, and the human being arose.

This is a very short outline of the development of our Earth in pre-geological times. The subject needs careful and detailed study before it can be truly assessed.

When Rudolf Steiner in his note speaks of the expiring and extinguishing of the collective beings of plant, animal and man he means that from the all-pervading, living plant substance the individual plants begin to emerge. The same applies to the undifferentiated living and feeling animal substance from which individual animals begin to arise.

In the case of man, Rudolf Steiner does not use the word expire or die away (erstirbt), but extinguished (erlöscht). This suggests that in man what has been extinguished can also be rekindled. For it is the task of man to bring down to the Earth in full waking consciousness the essential reality in which he lives before birth and after death.

This passage in Rudolf Steiner's note recalls a saying from the fragments of Novalis: 'When a spirit dies, a man is born: when a man dies, a spirit is born'.

Man is 'a citizen of the universe and a hermit on earth' as Rudolf Steiner once said. Man is really in his true element before birth and after death. With earthly birth the human being who was formerly embedded in the whole universe is extinguished. But on the Earth he can try to regain in his consciousness what he experienced before birth and will experience after death. He can rekindle in himself his true being. Originally man was permanently embedded in the life of the whole universe. When the phenomenon of birth began – as man emerged from the floating existence similar to what he experiences today in the womb before birth – he appeared in solid physical form. At the same time the rocks and salts solidified out of the fluid Earth.

So we see there were conditions on Earth when certain groups of substances were really imbued with life. Substance was fluid, permeated with warmth, air and light. This was the primal albuminous atmosphere of which Rudolf Steiner speaks in his lectures *Mystery Knowledge and Mystery Centres*. The substances which later hardened into rock were in solution in this primal albumen. What was precipitated as mineral substance from the life of the world-plant-animal in its dying away became porphyry. What was precipitated from the all-embracing plant nature of the Earth became *laminated rock* (schists, slates, etc.). What was precipitated from the animal nature became *limestone*. Only after all these mineral substances were precipitated from the albumen and their true being had died, did the present-day forms of life appear in all their diversity.

Finally, with the appearance of man the *salts* were precipitated out of the 'universal fluid' (Weltenwasser), and at the same time the still soft rocks began to harden. A surviving remnant of this universal fluid, this original albuminous atmosphere, is the ocean with its high salt content. Another is the amniotic fluid in which man and animal float before birth – this fluid which contains salt, like seawater, as well as considerable sugar and albumen.

Animal and man have taken into themselves in their blood something of this primaeval albumen. The salt content of the blood resembles that of the sea. Here the higher organisms have retained something of the original living and sensitive substance from a past stage of the Earth development in order to transform it into the foundation of their feeling and conscious life.

In the last sentence of the first part of the above-mentioned quotation Rudolf Steiner speaks of 'sulphur in which the mineral is burnt and consumed.' This terminology is foreign to modern abstract thinking. The concept of 'sulphur' has little to do with the substance sulphur we know as a mineral. 'Sulphur' is an old alchemical term signifying fire or heat. This idea of sulphur referred to a process which took place in heat. One saw a living 'sulphur' for instance in the flowering of plants. Rudolf Steiner uses the term to indicate the living fire in which the minerals are consumed.

This implies that through a living heat process certain mineral substances i.e. silica, aluminium, magnesium, etc. began to separate out of the primaeval albumen as concrete substances with firm outline. This does not mean however that crystals appeared.

In this burning process the first beginning of our later minerals appeared, especially those found in granite and the greenstones. In other connections Rudolf Steiner has spoken of this burning-up process as an all-embracing flowering process of the whole Earth.

He speaks of a world of *mineral-plants* as a repetition of a previous stage of development of the Earth.

Later, as we go into detail, we shall see that the flower and plant nature of this 'burning-up' is still to be discovered pictorially in some rocks.

This mineral-plant-world is very ancient and existed already on the Old Moon stage of the Earth (see: *Outline of Occult Science).* Then it was still living. Now, in the repetition on our Earth, it dies and in so doing forms granite and greenstones, the firm foundations of our Earth.

3 The World of the Mineral-Plants and their 'Signature'

In the last chapter we spoke of a past stage of life on our Earth which came to its fullest expression during an earlier incarnation of the Earth which Rudolf Steiner in his *Occult Science* calls the Old Moon.

This Old Moon was a 'fluid body' which did not yet contain any hard minerals but was through-and-through rampant, living growth. Rudolf Steiner describes the mineral-plant world of the Old Moon as an inwardly proliferating, peat-bog like substance. This of course only gives some sort of an idea as our modern peat-bogs are not 'inwardly proliferating'; their only life is on the surface and the edges. These modern peat-bogs are an earthly phenomenon which only appeared later when the real life of the fluid Earth was confined to the surface and the dead substances had sunk to the bottom and later hardened into rock.

Another modern phenomenon that is perhaps nearer to the rampant growth of the Old Moon is the way a cancerous tumour grows in a human or animal body. Here, in a pathological way, we have the remarkable fact that cell growth is no longer controlled by the whole organism with its differentiated structure, but becomes self-willed and runs amok. What is significant is that a cancerous tumour often shows embryonic forms which belong to an earlier stage of development of the organism. In this disease therefore man, in a way, slips back into an earlier stage of development, he develops something of the Old Moon nature in his body and is no longer able to control growth into definite organs and tissues.

The rampant growth of the mineral-plant world of the Old Moon showed a distinct differentiation which came about through the fact that this Old Moon at the beginning of its development included what today circles the Earth as the Sun. It was a heavenly body that consisted of today's Sun, Moon and Earth. In this state the Old Moon *recapitulated*, to begin with, those earlier conditions of the Earth that we have called Old Saturn and Old Sun. During the recapitulation of the Old Saturn condition the life of the mineral-plant was permeated with warmth – was *flower-like*. When we see in the high mountains thick cushions covered in tiny blossoms emerging from the dry rocks and surrounded by flying insects we have a picture of the life processes of which Rudolf Steiner speaks. If one can picture myriads of such flower cushions interpenetrating each other luxuriantly, it gives one some idea of this remarkable life.

This flower quality of the life of the Old Moon which was a repetition of the Old Saturn development gradually became more tree-like. It did not however form branches and leaves but life-forms resembling the annual rings of our present day trees. Wood and root-like forms appeared which densified to a horn-like consistency and then dissolved and flowed away again. These fibrous and layered forms of life were not hard or clearly defined but were in a state of continual change or flux. This tree-like vegetative growth was a recapitulation of the Old Sun development of the Earth.

With these two recapitulations we are dealing only with mineral and plant processes and forms; there is nothing of an animal nature present yet. Then comes the separation of Sun and Moon, the Sun becomes an independent heavenly body around which the Old Moon revolves. This Old Moon now begins its own proper development and a part of the substance of the mineral-plant becomes endowed with feeling. The new kingdom of the plant-animal appears. But here too, life is an ever-flowing process, never taking on fixed forms. The Sun, acting on the Old Moon from outside, effects a sort of fertilisation in the life of the animal-plants.

In the whole of the life of the Old Moon there are no fixed permanent forms, only densifications to a horny consistency which soon dissolve and return to a state of flux. Towards the end of the Old Moon development, of which we have given a brief account in order to make the later stages of Earth development comprehensible, the two heavenly bodies of Sun and Old Moon reunite and pass away into a state of rest, a 'Pralaya'.

In the beginning of the Earth development all the previous stages of Old Saturn, Old Sun and Old Moon were recapitulated according to the biogenetic law, before the Earth development proper began. The 'Earth', which initially was a single heavenly body containing Sun and Moon, first went through a condition of pure warmth which was characteristic of Old Saturn, progressed to the air-light body of Old Sun, and finally to the fluid body of the Old Moon. These three stages were also separated by rest periods. After the third rest period there began the fourth part of the Earth development which had as its goal the densification, step by step, into three dimensional, physical, material existence of all that had gone before. Within this fourth stage of Earth evolution, the Old Saturn, Old Sun and Old Moon conditions were repeated with enhancement and metamorphosis. These repetitions were also separated by rest periods. After the third rest period the true physical development set in and led to the solid crystalline state of the present time after again recapitulating the Old Saturn heat condition, the light-air condition of the Old Sun and the fluid Old Moon condition.

Rudolf Steiner describes this threefold repetition in his book *Cosmic Memory* in considerable detail. We have outlined the essentials in order to give some idea of the extraordinary complexity of the development which preceded the physical material emergence of the mineral kingdom.

In this threefold recapitulation on the Earth, the world of the mineral-plants reappeared passing through all stages of densification from pure warmth, through the gaseous and fluid states, to the fourth condition – the solid. In this solid state of the rock that formed, we can distinguish older and younger 'igneous' formations of granite and greenstones, and the formation of trachyte and basalt. In the older formations of granite and greenstones we see the remains of the first recapitulation. In the younger formation of the same rocks we have the remains of the second recapitulation, and in the trachytes and basalts the remains of the third recapitulation.

In all these recapitulations the stages progress from heat to fluid. The whole development is closely linked with the development of the human organism. During the third recapitulation, the 'mineral' condenses to the vaporous-fluid-gelatinous condition which Rudolf Steiner describes as the Lemurian albuminous atmosphere.

In picturing these progressive solidifications and rarifications one must remember that they also have a spatial aspect. Rudolf Steiner has said that, in its pre-physical development, the Earth extended far into planetary space and through a gradual rhythmic shrinking process acquired its present size. During these processes life, which to begin with filled the whole body of the Earth, withdrew more and more to the periphery of the planet where there was now an atmosphere in which it could develop further.

Such was the Lemurian albuminous atmosphere. In it were contained all life processes and substances in an undifferentiated state. It was the sheath which nourished, illuminated and warmed the living Earth. The world-plant-animal, together with the essential being of plant, animal and man, lived in it and constituted the life of the Earth.

The oldest life of this living albuminous atmosphere was the world of the mineral-plants which was brought over from the Old Moon in the recapitulations described above, and now came to an end in a 'grandiose combustion'. This process of combustion described by Rudolf Steiner in the note quoted in chapter two was, in another context, characterized by him as a sort of flowering process. In order to understand this we have to look at processes which take place in the flowers of our present day plants. What we mean is seen most clearly in annual plants. Let us take plants such as chamomile or marigold (Calendula). In their flowers, colouring matter and strong scent are formed. The plant which, before it flowers, is all growth and development, suddenly calls a halt in order to make room for a process which one can only compare with a burning away – a flaming and fading away (disintegrating). What arises from this decomposition are essential oils. These are inflammable and burn with brilliant coloured flames.

It is noteworthy that every blossom has a considerably higher temperature than its surroundings. Here we have really an organic process of combustion. The process does not go as far as mineral combustion. We smell no smoke, only the scent of the flower, we see no flame, only the colour of the flowers, we see no ash, instead we see the seeds which result from this remarkable plant-combustion.

The life of the mineral-plant was a pure blossoming process – what Rudolf Steiner in his note calls 'the mineral being consumed in sulphur'. No leaves nor roots belong to this blossoming process. The mineral-plant was an organism that floated in the living, nutritive substance of the albuminous atmosphere and was open to the forces of the surrounding cosmos and the Sun. Into this life of the mineral-plant there dripped and streamed down from the surrounding spheres of the Earth that which later densified as quartz and silicious minerals. It caused this life to densify into flowing forms which came and then went again, allowing the more or less formed silicious substances to fall out. These processes were connected with light phenomena – with a glittering and 'greening' which formed and dissolved. Rudolf Steiner describes this life of the mineral-plant in his lectures *Mystery Knowledge and Mystery Centres* (November/December 1923) where he shows that the knowledge of this life on the Earth in times long past was the content and teaching in the pre-Christian temples and mystery centres in Greece and Ireland. He also points out that in the present day rocks there are forms which are something like memories of these old times. He says: 'Anyone who has looked round a little on nature knows quite well that something like distinctive marks of an ancient time are to be found in the mineral world today. We find stones, we take them into our hands, we look attentively at them, and we find they have within them something like the outline of a plant-form.'

Before we go into this any further and describe the relevant phenomena in the rocks, we would mention that Rudolf Steiner, in very early lectures in 1907 and later in his lectures to the workmen at the Goetheanum and at teachers' meetings in the Waldorf School, pointed out that the minerals – quartz, mica and felspar which make up granite and gneiss – are to be seen as the remains of the flowering process of these mineral-plants. The laminate structure of the glittering mica suggests bracts; and the felspar, pistil-like forms. Quartz is to be seen as the filling between these forms. Hornblende which appears in these old rocks is also to be seen as a metamorphosis of these bract-like forms.

As Rudolf Steiner speaks of these 'memories' of old times we might add that mica is often found in rosette-like aggregates – 'mica roses' (See plates 5 and 6). This is an unusual arrangement of the individual mica crystals which normally form small hexagonal prisms.

The author has found such mica roses in the Fichtelgebirge and the Black Forest. They are sometimes embedded in the solid rock but can also appear in clefts and cavities as free standing groups of crystals. The latter especially give a strong impression of a plant-like, calyx-like grouping.

Near Hermannsschlag in Moravia one finds ball structures (Glimmerkugeln) in which brown mica alternates with fibrous actinolite and nephrite (jade). Near Rozena in Moravia one finds green flaky prisms of zinnwaldite (lithium mica) with a sharp hexagonal cleavage which are crowned with flaky rays of crystals. Spherical-curved, laminated mica is to be found near Skogböle in Finland. From these few examples, which one still finds described in the old classical textbooks, one can see what is meant. There may be many more forms of mica which are reminiscent of flower bracts. But these phenomena were considered accidental, they were ignored as unimportant and were not described. It will be a task for the future to collect and describe such phenomena which Rudolf Steiner called 'memories' (Merkzeichen) so that we may obtain concrete evidence for the findings of his spiritual science.

These 'memories' are even more evident in the minerals which compose greenstones and schists. The principle minerals are augite (pyroxine) and hornblende (amphibole) (Plates 7 and 8). These illustrate that stage of the mineral-plant development which we described as wood and tree-like, which indicates the repetition of the Old Sun condition.

At the end of this development which leads from the flower-like to the wood and tree-like mineral-plant world, we have serpentine which, in its duality of foliaceous and fibrous serpentine, forms the bridge between the laminated mica and the fibrous (i.e. long, needle-like) minerals of the pyroxine and amphibole groups.

This transition follows two great series of rocks; from granite through gneiss to crystalline schists (which include part of the schistose greenstones) on the one hand; and from granite through syenite and diorite to diabase. At the end of both series we have serpentine which, as foliated serpentine (antigorite), leads through chlorite to mica, and as fibrous serpentine (chrysotile), (Plate 9) to hornblende.

In these groups of rock which occur in nature in all variations together, we see the development in impressive metamorphosis from the granular form-principle of granite to the foliated and fibrous form-principle which appears later in the plant world. The point of departure of this development is mica – this pale image of the primal leaf which has been left behind as a 'memory' of the beginning of life.

We shall see later that this original formative principle of mica is still active today not only in the creation of the plant world but also in the formation of the living, plant-bearing soil.

The threefold repetitions of previous conditions on their long path through many metamorphoses, developing the foliaceous and fibrous form-principles, have left behind a multitude of forms which bear witness to these metamorphoses. From the relative simplicity of granite, a wealth of rocks and minerals has developed which is comparable to the richness of our present plant world.

In the widespread wood-grained structures of the crystalline schists and greenstones, which appear embedded in the Earth like annual rings of giant trees, there are innumerable little 'memories' hidden away. The most obvious is asbestos (Plate 9) this densely fibrous mineral that can be spun and woven like plant fibres. We find bissolith, mountain flax, mountain leather and mountain cork – all minerals related to asbestos – minerals which are pliable like leather and which float on water like cork.

Then there is chrysotile – green, densely fibrous layers embedded in gabbro-greenstone, the brown glittering fibres of bronzite, the green diallage and diopside (Plate 7), green pistacite and nephrite, all of which when embedded in steatite (soapstone), schist or quartz, form grass and plant-like structures. On the cleavage surfaces of mica schists one finds delicate crystals of nephrite and hornblende in enchanting forms of fans and sprays (Garbenschiefer) (Plates 8 and 9) and patterns resembling leaf veins.

Perhaps the most beautiful of what Rudolf Steiner calls 'memories' is to be found in the exquisite green seaweed forms of the moss-agate (Plate 10). Here the bluish translucent threads of hornblende-asbestos or the micaceous chlorite are embedded in a gel-like chalcedony. If one cuts this mineral into thin slices only a few millimetres thick and illuminates them from behind, one seems to see green plant forms floating like seaweed in the ocean.

It is remarkable that most of the minerals that we have called 'memories' are green. When we look into the substances of these 'memories' it becomes apparent that the 'plant-nature' of their origin is also expressed. The main constituent of all the above minerals including their parent rocks (crystalline schists, greenstones, gabbro, serpentine, etc.) is magnesium silicate, and the green colour is caused by iron silicate. So we have three substances which are essential for our present-day plant world. First magnesium which, in the chlorophyll of green plants, takes the place of iron in the haemoglobin of human blood; then iron itself that must be present in the soil and air surrounding the plant so that the chlorophyll in the plant can be produced. And finally silica which not only enables the plant to absorb the light but is built into its rigid form and structure.

From the above it is clear that the old mineral-plant world which originated in the past life-stages of the Earth, still had within it the essence of present minerals. In order that the living plant could develop further as an entity and rise to higher stages of existence, this life had to die. Its corpse we find in a wealth of remarkable rocks and minerals which as 'memories' still speak to us in a language we can understand.

4 The 'World-Plant-Animal' and its 'Signature'

The concept of the plant-animal is something which to our modern way of thinking is as evasive as the concept of the mineral-plant which we attempted to outline in the previous chapter. If we want to get closer to this concept we have to remember that Rudolf Steiner himself characterized this all-embracing, undifferentiated state of development of life as 'world-plant-animal'. This suggests a life process that encompasses the whole Earth, and not differentiated kingdoms of nature existing side by side.

We already mentioned that this life process reached its climax in the Old Moon evolution of the Earth and then reappeared in a modified form during the recapitulations at the beginning of the present Earth development.

What really happened in this life process of the plant-animal cannot be put into modern scientific concepts. This is due to the fact that present-day nature with its differentiated forms simply did not exist.

One might be prompted to ask: What has become of the plant-animal in the course of evolution? What sort of signature has it left behind and where are we to look for the present-day forms of life which have developed from it?

In the mineral-plant we could recognise the two life processes which we find today in blossoms and in trees. We showed that from the granular form principle of granite, foliaceous and fibrous ones developed as found in foliaceous and fibrous serpentine. If we follow the corresponding transformation of the plant-animal, it becomes far more complicated. If one wants to get anywhere near the right ideas, one must picture the general connection of the present-day plant-world with the insect-world on the one hand, and on the other hand, with the living soil in which it grows. One must not forget that there is not only an interdependence between insect-world and flower, as well as other parts of the plant which are above the earth's surface, but that the larvae of a great many insects are equally necessary for the proper life of the soil in which the plant roots live.

If one can view the interaction and interdependence of the insect-world, the plant-world and the soil as a living unity, one can say: This is what has developed in the course of time from the old 'world-plant-animal'.

The interaction of these three realms which embrace the whole Earth is now, as formerly, the foundation of all animal and human existence and has its root and origin in the world-plant-animal. All the phenomena of what is called symbiosis in the plant and lower animal kingdoms – in earth, water and air, even the intestinal flora and fauna in man and animal – have their origin in the world-plant-animal. They are its remnants and traces – transformed.

Here it is possible to add something which still has to do with the concept 'mineral-plant'. Up till now we have considered those forms which have resulted from the old life processes of the 'mineral-plant' and the 'plant-animal'. For the former these are the flowering processes of the higher plant and the trunk and root formation of our trees. For the latter it is what we have described above as the living connections of insects, plants and soil. There also are the rocks which must be considered as something that has been cast off and left behind. There is though, in every development, a third possibility; the old life processes may be slightly transformed and projected into the present. A short summary can make this clear.

The mineral-plant:

Has *left behind as rocks:* the granites, greenstones, schists, etc. Has *developed further* into trees and the flowers of the higher plants.

The plant-animal:

Has *left behind as rocks:* the porphyries and such like. Has *developed further* into the great living community of insects, plants and soil.

The third possibility lies between these two. Thus it has to be a process lying between dead rock and the life processes of plant and insect. We must seek it in the soil where the plant roots are and the insects spend their larval stage.

What takes place within these really very obscure life processes of humus formation deserves to be considered as yet another kingdom of nature, and although this subterranean realm is closely linked with mineral, plant and insect, yet it has its own laws of life. They are the primaeval laws of the 'plant-animal' that belonged to the Old Moon condition of the Earth and which, transformed through recapitulations, has now become the soil – the foundation of life. Our present-day soil corresponds to the living atmosphere of the 'Moon-Earth' in the early periods of our present Earth.

When one considers the formation of humus, then three fundamental processes become obvious. The first one, apparently entirely mineral, takes place in the weathering of rocks. The second process takes place in the dead remains of plants and animals. This second process is brought about through the agency of lower plant and animal organisms such as algae, bacteria, fungi and protozoa. The third process is the unification and compounding of the results of the first and second processes through insect larvae* and earthworms.

The first, the mineral process, is to begin with a dissolution and disintegration. The constituents of the rocks – quartz, mica, felspar, hornblende – decompose through weathering by water, air, heat and cold into their component parts. The resultant end-products are finely dispersed silica, aluminium oxide, magnesium and iron oxides, and the alkaline salts of calcium, sodium and potash. Through the action of the all-pervading water these substances, especially silica, aluminium, magnesium and iron oxides, become colloids which are extraordinarily unstable and will easily revert to the crystalline state.

Under certain conditions found today almost exclusively in virgin forests or steppe-soils, there now begins the reconstituting part of this apparently mineral process. From the above mentioned colloidal (gelatinous) substances quite new minerals emerge. Under these apparently entirely mineral conditions, really only mineral salts should be formed, yet the remarkable fact is that very complex compounds of these substances occur – the secondary clay minerals. These elaborately formed, secondary clay minerals have the inner (molecular) structure of mica. But these mica-like minerals are not crystalline like true mica but are colloidal like the original substances

*including the many insects which retain larval forms throughout their life such as wood-lice and centipedes. (translators' note).

resulting from weathering. This means primarily that they have an extraordinary capacity to retain water.

The reconstituting of minerals resulting in entirely new ones (of gelatinous, mica structure) from the products of weathering, is a true life process of the Earth because it does not follow ordinary chemical laws (if so the result would be salts) but the laws of the life of the mineral-plant world. This is indicated by the fact that mica-like and therefore foliaceous structures arise, and that besides clay and silica, magnesium and iron play a major role in the formation of these new minerals.

Thus we have attempted to sketch the third aspect of the mineral-plant existing at the present-day as a living process between what has been left behind as rocks and what has developed further as flower and tree formations.

The other aspect of the life of the mineral-plant is the second fundamental process of humus formation described above. It is the transformation of plant and animal residues in the organic part of humus. This organic part of humus which gives our soils their dark colour is a remarkable living substance. It is closely related to tannin which is found in the bark and wood of our trees and in the roots of many plants. If one examines the chemical constitution of this substance, it is interesting to find that it is also related to certain substances found in the scents and essential oils of many plants. Thus we have the remarkable fact that in both tannin and humus are to be found the transformed essence of the flowering process. This fact throws light on Rudolf Steiner's observation that the life of the mineral-plant was a 'blossoming' or 'blooming' life. However this blooming is under the surface of the earth, therefore it is not essential oils that are produced, but the organic part of humus. A whole host of minute organisms is engaged in producing the dark humus out of the remains of dead plants, wood and cellulose, as well as plant and animal proteins. There is a very well-organised division of labour between algae, bacteria and fungi receiving and handing on substances at the appropriate stages so that finally, out of the most diverse plant and animal remains, there emerges a uniform humus.

In recent decades all these processes have been thoroughly investigated especially by Russian research workers. German scientists like Laatsch and others have taken up these studies and have reached important conclusions concerning the creation and structure of the organic part of humus. Of particular importance is the fact that humus has a resemblance to mica, like the secondary clay minerals previously described, and the physical constitution is colloidal, like the clay minerals.

One cannot overrate the strange fact that here, in miniature in nature, these two processes take place. The one, apparently purely mineral, leads to the formation of new secondary clay minerals. The other, through the digestive process of lower organisms, produces humus. Both substances are composed, like mica, of minute flakes, and are gelatinous – colloidal.

Each of these two substances is, because of its colloidal structure, extraordinarily unstable and very sensitive to changing conditions of moisture, acidity or heat in the soil. Through even slight changes both can loose their colloidal state and become crystalline. In this case the secondary clay minerals would become stone and the humus gradually peat or coal. This means that they would slip out of reach of the life forces, for these can be active and formative only on substances in the colloidal state.

These two processes, the formation of clay minerals and of humus as living, colloidal substances, can be very easily disturbed or even stopped by faulty cultivation – lowering of the water table (drainage), clean felling of woodlands, monocultures and above all by fertilising with chemicals. Salts destroy the colloidal state of the soils because colloids are flockulated by salts. They can then no longer be a vehicle for life forces.

What works on the clay minerals and the humus while they are in the colloidal stage are the life forces of the plant-animal. It is of the greatest significance that both these substances are eaten by the soil fauna. The organisms are – according to the nature of the soil (woodland, meadow or arable) – wood-lice, mites, springtails, insect larvae and above all – earthworms.

These creatures devour the earth, they seek places where the two substances are produced. Their life consists in compounding the two substances by means of their digestive and metabolic processes into stable, ripe humus. They are the great artisans who, in their diminutive intestines, achieve the complete amalgamation of the minute particles of clay minerals and humus acids, combining these with nitrogen, calcium, etc. so that a substance is created that plant roots can take up directly as food.

This end product – stable humus – is still colloidal. It has a real though simple life of its own which is expressed in its extraordinary capacity for retaining water and replenishing from the surroundings what the plants have extracted. If these substances in the soil are husbanded they are a hidden treasure which, like the widow's cruse, can never fail.

With this we have described the third process in the formation of humus. We have seen that the first stage of these three processes takes place in the purely mineral realm (the secondary clay minerals). The second involves mainly the plant world (the humus acids) and in the third process the 'animal' takes over and creates something new out of the mineral and vegetable material that is neither mineral, vegetable nor animal.

The life of this humus is a synthesis of the mineral-plant and the plant-animal in close conjunction with the individual living organisms in the soil.

Now we can complete our brief summary:

The mineral-plant:

1) Has *left behind as rocks:* the granites, greenstones, schists, etc.
2) *Is active today* in the formation of the secondary clay minerals and organic humus acids.
3) Has *developed further* into trees and the flowers of the higher plants.

The plant-animal:

1) Has *left behind as rocks:* the porphyries and such like.
2) *Is active today* in the joining together of the building stones of humus (clay minerals and humus acids).
3) Has *developed further* into the great living community of insects, plants and soil.

This presentation of the living connection between rocks and the kingdoms of nature as they are today in all their variety, is not only of importance for our study of rocks. If one can really grasp these relationships one has the necessary basis for understanding the practical applications of the agricultural system arising from Rudolf Steiner's researches.

A future agriculture and forestry that does not take into account these relationships will contribute to the decline of the nutritive value of our plants, and our forests and crops will more and more become prey to pests. The pest problem is a revolt by nature against the *disturbed harmony* between soil, plant and insect. The disturbing of this harmony started when fertilising with salts became general practice. The result was the upsetting of the natural life of the soil.

The laws that govern the life of the mineral-plant and plant-animal are fundamental laws of the whole earth and the kingdoms of nature. These laws originated in the distant past of the Earth and, with many transformations, have persisted to the present-day. We have come to know them in the interaction of mineral, plant and insect. In this the insect appears as a guardian of this harmony. The *winged forms* of these real plant-animals, headed by the bees, control this harmony by bringing together *above the earth* the pollen born of the air and sun with the ovary born of the watery moon forces. The resulting seed can however germinate only when, *in* the earth, the *insect larvae* maintain harmony between the sun-related substances (silica, magnesium) of the clay minerals and the dark, moon-related substance – humus.

The second process is even more important. It is the *fertilisation of the earth*. What the bee is above the earth, the earthworm is within the earth. The earthworm is king among all those who fructify the earth because it has remained a 'larva' and therefore renounced the chance to become a 'butterfly'. Its red blood, containing iron, shows that it has developed far in advance of related forms. It is *the* representative of the old world-plant-animal because it knows best how to recreate as humus the life of the mineral-plant which has fallen apart into mineral and plants.

*

Before we proceed to describe those parts and processes of the rocks which are connected with the old life-stages of the Earth let us consider the following:

In the processes connected with humus formation, besides bacteria, fungi and algae (the simplest plant organisms), there is a group of lower animals which plays an important role – the protozoa. These consist of rhizopods with their sub-groups – radiolaria, amoebae, foraminifera, etc. There are also the flagellates some of which, like plants, form chlorophyll or have a cellulose shell. Tripanosoma (sleeping sickness) and spirochaete (syphyllis) which are present in human and animal diseases are flagellates. The last group of these unicellular animalculae is the infusoria. Among these creatures are some who live in the soil and contribute to the making of humus, and some who live as parasites in higher organisms. These lower animals are still extraordinarily plant-like in their metabolism as the appearance of chlorophyll and cellulose shows. Some of them however are also closely related to the mineral, for example the radiolaria build their shells of silica and the foraminifera of calcium. Here are organisms which form a bridge from the mineral-plant to the plant-animal.

Thus in protozoa we have organisms that clearly point to the mineral-plant and to the plant animal. They are metamorphoses of those ancient forms of life on Earth. All symbioses (such as between root and root-fungi), which contribute constructively to the great household of nature, are also true metamorphoses of these old forms. Parasites, however, have not followed this harmonious development. They have remained behind and eke out an existence at the expense of higher forms of life.

Another transformation of the world-plant-animal is to be found in the fresh and salt-water animals known as coelanterata. This is a group of animals that even modern zoology calls plant-animals since they are for the most part permanently attached and outwardly resemble plants. To these belong sponges, polyps, jelly-fish and corals. When, in an aquarium, one contemplates these oldest animals, one is transported back to a time when, from the undifferentiated life of the whole Earth, the first forms emerged, fleeting and floating, colourful and transparent. In the oceans these first attempts at creation remain for us to see in all their flower-like beauty. They manifest a capacity to produce forms which far outstrip human art. Any one who wishes to pursue this further should study Haeckel's 'Kunstformen der Natur'. (Art Forms in Nature.)

The traces and 'memories' that the life of the world-plant-animal has left in the rocks are very diverse. In order to understand better the nature of these traces, we must take a look at the special characteristics of the 'animal-nature' and how it manifests itself in the lower animals and also in the embryonic development of higher animals. Primal forms of life are spherical. One has but to think of a seed, a birds egg or the ova of the higher animals which we call eggs in spite of their small size.

In every such ovum or seed the organism returns to the primal beginning, to where the individual forms of nature began to emerge out of the living world-body. From this primal beginning of the seed, the plant-nature develops into the planar surfaces of leaves and the linear formation of the stalks and inner vascular structures. The essence of plant life is the flat leaf from which it derives its nourishment. With the appearance of the flower a new element enters that goes beyond the leaf. What happens is that a number of leaves unite to form a more or less distinct cavity – the ovary – in which the seeds are formed. The creating of such cavities is not a typical plant process. It happens because in flowering and fruit formation something of a higher nature acts on the plant; something of an animal nature. This principle is visible in the relation of flower to insect. Flower and insect belong together because they both originate in the plant-animal. This animal or astral principle expresses itself in the animal world through the forming of cavities in the body, such as bladder, stomach, uterus, etc. In the embryonic development of the animals the building of cavities by invagination can be followed in all its stages. The characteristic of the animal is thus its capacity to form organs, which are cavities, by invagination and evagination. By this, certain processes of the environment are drawn into the interior of the body. Whereas the plant is able to take in many substances from the air through the surfaces of its leaves, the animal has drawn an outer surface into its body and formed the passage of the stomach and intestines. Only by means of these 'in-folded leaves' can it digest and make use of its food.

Another formative principle in animals is expressed in the differentiation of the inner organs as regards structure and function. The inner organs, which are all to a certain extent cavities, are all composed of cells. But these cells are modified into liver cells in the liver, kidney cells in the kidneys, lung cells in the lungs, and so on. All the cells are formed after the same pattern, but their shapes and functions differ fundamentally in the individual organs. By invagination the animal creates both the cavity *and* the particular inner structure and function of the organ. In this way the embryo, after initial simple growth, differentiates itself into a inner world of organs which in their functions support each other.

This forming of cavities and differentiating (one might even call it individualising) has a counter-picture in the world of rocks. To study it in its first beginnings we must walk over the old mountains of crystalline rock that form the backbone of the Earth. In Europe they are for instance the Bohemian Massif, the Bavarian Forest, Fichtelgebirge, Odenwald and the Massif Centrale of the Auvergne. In the huge quarries of these mountains one can often observe that in the uniform granite mass there appear 'veins' and 'nests' in which the individual components of the granite are enlarged. The uniform granulation of the granite composed of millimetre sized crystals of quartz, mica and felspar is suddenly interrupted. In the neighbourhood of these streaks and nests the granules become bigger and bigger until they fill in a seam or emerge as free standing crystals in the cavity. This formation in a crystalline rock is called *pegmatite*. It does not occur in faults and clefts caused by stresses and mountain upheavals, but is found in the solid rock masses. Therefore one must not confuse this phenomenon with the crystal clefts and caves which occur in the younger crystalline rocks such as the Alps. The crystals of pegmatite can vary from the minute to the gigantic and they are not always associated with cavities. The largest pegmatites can yield sheets of mica of more than a metre in diameter, together with enormous blocks of quartz, rose quartz and felspar. In the Urals there is said to be a felspar quarry in a pegmatite consisting of one single crystal.

The forming of pegmatite was a process that worked from within on the gelatinous fluid mass of rock-in-the-making. The volume of this gelatinous mass was many hundred times that of the later solidified rock. The formation of enormous crystals was *not* in any way associated with existing cavities but it occured *within* the gelatinous mass. If one finds cavities in which crystals have grown, one must assume that originally these were much larger, and with the solidifying and drying out of the rock they have shrunk tremendously. It is characteristic of dense pegmatite, where huge crystals of quartz, mica and felspar are packed as closely as in fine grained granite, that larger forms arise than in the pegmatite veins which wind snake-like through the rock.

Besides the crystals which constitute the major part of granite – quartz, mica and felspar – there appear in the cavities of pegmatite a number of other crystalline minerals. These are mainly *precious stones* topaz, beryl, tourmaline, and precious corundums such as saphires and rubies. At first sight it appears that these minerals are only present in the cavities but on closer observation of the surrounding fine-grained granite it is found that these precious minerals are distributed in minute particles in the rock mass. The large crystals of the above stones appear to be drawn into the pegmatite cavities out of the granite mass.*

Two phenomena – the forming of cavities and the differentiating (individualising) of the cells of the organs which we have found associated with the forms of flowers in plants and organs in animals – we have now met with in the world or rocks. If we call to mind Rudolf Steiner's remark that the forming of granite goes back to the flowering process of the old mineral-plant world, then these phenomena can be understood as the first *traces* of the influence of an animal (astral) principle, higher than that of the plant. In the 'inwardly proliferating, peat-bog mass' of this mineral-plant world there appeared cavities in which the individualising principle of crystal formation could work. The crystals which developed in the organ-like cavities are inner *sense organs of the Earth.*

Along with this pegmatite structure occurs another which is the very opposite. In granite and other granular rocks one finds veins filled with a very fine-grained material. If this material resembles very fine-grained, light-coloured granite it is called aplite. If it is dark and rich in iron (hornblende and augite) it is lamprophyre. Aplite is nothing more than a granite almost entirely lacking in mica, and lamprophyre is a rock that consists mainly of felspar and hornblende as well as augite with practically no quartz.

There is an enormous number of these rocks named after their

*More about precious stones is to be found in the author's book 'Kleine Edelsteinkunde', second edition, Stuttgart 1965.

local incidence and variations, and there is no point in enumerating them all. What matters is to grasp the processes which lead to their formation. If, starting from pegmatite together with aplite and lamprophyre, one can follow up the whole process and its extention and branching out into all the crystalline rocks including gneiss and crystalline schists, one can understand that here one is in the presence of a mighty process which one can designate as porphyry-forming. The differentiation and endless multiplicity of these rocks is a clear sign that instead of the life-processes that have formed the simple granite, we have here to do with other and more diverse life forces. The veins, streaks and seams of these aplites and lamprophyres wind through the solid rock mass like the limbs of a gigantic animal or the branches of a mighty tree. They gradually break up, cleave and transform the primal unity into multiplicity; a multiplicity not only of substance but also of form. It begins with pegmatite and culminates in the amigdaloidal melaphyres, of which more later.

What happened exactly? The two processes, one producing cavities and one 'individualising' certain minerals in these cavities, acted upon the gelatinous mass of what later became rock, and differentiated it in an 'inwardly proliferating' growth process. If these phenomena were looked upon with impartiality it would never occur to anyone to imagine that these masses were ever fiery and molten. The immediate impression is of something that has 'grown' in the true sense of the word. We find this preserved in the language of unsophisticated miners and quarrymen.

The porphyry-process in all its diversity is the expression of the forming-power of the world-plant-animal as Rudolf Steiner calls it. It acts on nearly all the crystalline rocks, and its effect is more evident in the *forms* it produces than in the actual substance. What are the characteristics of this porphyry structure or form? Let us call to mind once again the pegmatite with its individual crystals and then both aplite and lamprophyre with their shapeless fine-grained mass. On the one hand we have the strongest formative force, and on the other a setting-in of formlessness. At first both parts of the process run separately. Starting from pegmatite the great granite masses develop towards porphyritic granite in which quartz and mica appear as small crystals and felspar as larger crystals. From aplite and lamprophyre develop the multitude of subsidiary types too numerous to name. Quartz and felspar dominate in aplite, felspar and hornblende in lamprophyre, mica being almost absent. When mica disappears it is always a sign that the plant nature retreats and an animal-plant element takes over in hornblende.

Both streams – commencing from pegmatite and from aplite/lamprophyre – can be described as porphyritic, but the true porphyry appears where both streams re-unite and form a rock that contains large, well-formed, felspar crystals embedded in a fine-grained or even glassy ground mass (Plate 2). Individual forms appear in the more or less formless medium. In this true porphyry mica has well-nigh disappeared. Granite has ceased to exist and has been transformed by the porphyry-process. The felspars – the most conspicuous crystals of the porphyries – are that part of granite which, from Rudolf Steiner's descriptions, we can recognise as the pistil in the flowering process of the mineral-plant. In felspar, however, the flowering principle approaches so near the 'animal-nature' that, in its chemical composition, it has an alkaline calcareous part. In his Agricultural Course Rudolf Steiner explains that the alkalis – potash, sodium and calcium – are closely related. So these felspar crystals in porphyry are the sign of the world-plant-animal.

In another line of development which proceeds from granite, globular or spherical forms are enclosed in the pegmatite ground mass. Continuing along this line we find orbicular diorite (Plates 3 and 4), orbicular norite (dark rocks consisting of calcium, felspar, augite, diopside and olivine) and orbicular gabbro. These spherical nodules which we find embedded in the rock are most singular. The core of the nodules – very evident in the polished cross section – sometimes consists of a foreign substance or more often of crystals or groups of crystals. In such a nodule the felspar can radiate from the centre and be surrounded by a concentric ring of mica. Light bands of felspar can alternate with the dark mica. An essentially organic picture emerges from the polished cross section. If we remember that the felspar was the pistil in the 'flower' of the old mineral-plant-world and mica the calyx, we see in such a cross section the picture of a flower with the pistil (felspar) in the centre and the calyx (mica) arranged concentrically around it. The rhythmic repetition in light and dark layers gives us the phenomenon of 'interpenetrating growth' of which Rudolf Steiner speaks. It is also expressed in the spherical shape. The 'flower' was so to speak enclosed all round and 'grew' within itself.

These orbicular rocks are reminiscent of the later agates. In fact in the subsequent development of these processes, we find amigdaloidal melaphyre – the matrix of the agate.*

Melaphyres are dark coloured rocks which resemble basalt and are composed mainly of minerals closely allied to hornblende and augite, with some felspar. Mica and quartz are absent. In the amigdaloidal melaphyre one finds those marvellous creations called amigdaloidal agates. The original 'almond-shaped' cavities are either completely filled with the finest layers of silica, or into the remaining hollow of an imcompletely filled 'almond' there spring forth the most wonderful crystals of amethyst and other minerals. Here we have again the principle of cavity formation which, at the beginning of the porphyry process, we found in pegmatite. If one cuts through one of these amigdaloidal agates one has immediately the impression of an organic form. Fine layers of calcedony and opal alternate as many as 7000 times within one centimeter. Every 'almond' has a 'blow-hole' through which the silicious substance was sucked into these 'hollow organs'. Such agates are pictures that, through their silica-fillings in fine layers like annular rings, speak of the *plant* world, and through their 'organ' shape of the *animal* forming forces. It is the last sign and 'memory' of the world-plant-animal. (More regarding agates is to be found in the author's 'Kleine Edelsteinkunde.)

*The orbicular rocks are relatively rare. Orbicular granites occur in Sardinia, Argentina and Finland, orbicular diorite in Corsica, orbicular norite in Norway and orbicular gabbro in Sweden and California.

5 *The 'Plant-Nature' and Schists**

'In schists', says Rudolf Steiner in his note-book (see chapter 2 p. 26) 'the plant-being expires'. What is meant here by schist is not a single rock but a process which formed those gelatinous mineral substances which have fallen away from life and died. As the porphyry process forms the separate crystals and cavities, so schist shows, in general, a layered structure. If we say that the porphyry process begins in granite with pegmatite, we can say that the schistose process begins with gneiss (Plate 1) which often lies immediately on top of granite. We have here the same phenomenon as in pegmatite. Gneiss consists of the same minerals as the nearby granite – namely quartz, felspar and mica (or hornblende) – but one no longer has a uniform granular rock, for these minerals begin to separate into layers.

On closer inspection it can be seen that it is mica or hornblende which are responsible for the layering. In gneiss, mica no longer appears in 'books' of tiny individual leaves, but the mica leaves lie singly in thin layers between quartz and felspar. Hornblende, which in granite forms blunt granular crystals, makes in gneiss sheaves and fans of long bladed crystals and lies flat in the same direction as the layers in the rock (Plate 8). It is clear that mica and hornblende are the two minerals which govern the structure of the whole rock. Quartz and felspar have to adapt themselves to the lie of these minerals.

We have pointed out earlier that mica and hornblende are associated with the petal or leafy formation of the old mineral-plant world. If we can see that these two minerals appear to govern the structure of the layered rock, then we may conclude that the forming

*The German 'Schiefer' covers all schists, slates and shales. The translators have endeavoured throughout to use the most appropriate word since there is no collective English term.

forces come mainly from the plant nature. In the true mica and hornblende schists and in phyllites this is even clearer.

As with the porphyry process, so here we have an extraordinary diversity of rocks, but they all show a schistose and layered structure. Sometimes the porphyry and schistose processes overlap and the outcome is a porphyritic schist or a schistose porphyry.

As the development of the schist/shale series progressed, the process became more and more dominant, so that at a certain epoch large areas of the Earth were covered with its enormously thick layers. In the Silurian and Devonian series slates predominate; but the process no longer produces crystalline schists. Slates are predominantly dark, fine-grained or flaky rocks which can be split into thin sheets. Roofing slates and writing slates are made from them. In Germany they are found mainly in Thuringia and the Rhenish Schiefergebirge, in the Hartz Mountains and the Fichtelgebirge. Slate consists of about 75% aluminium silicate which is flaky like mica. It also contains a great number of other minerals in minute grains and crystals, for the most part they are chemically identical to minerals that are also found in granite or the other older rocks, but their structure is entirely different.

When one surveys the world-embracing process of slate formation one wonders how this came about. It is extraordinarily tempting to conclude that these slates originated only by the external circumstances of disintegration of the older rocks and their re-forming from sediment falling out of mighty streams of water. It is contradicted by the simple fact that this slate-process was not a localised phenomenon but continued for enormous periods of time and took place over the whole Earth.

Rudolf Steiner says that in this forming of slate the 'plant-nature' dies. We find two important pieces of evidence in support of this. One is the fact that at the end of the great slate period the Coal Measures appear, and the other is that slate shows a decidedly organic structure.

Whosoever is familiar with the schists and slates must be struck by the endless wealth of forms that remind one of the wood of our tree trunks. What we have already noticed as woodlike structure in the greenstones and greenstone minerals, appears again here in gigantic forms. Whole mountains of gneiss (Plate 1) or crystalline schist might almost be great masses of petrified wood. Magnificent folds of light and dark layers of quartz, mica, hornblende and felspar look as if the maleable, soft primal substance of what later became rock, was formed by a sort of rhythmic growth. One must not however imagine that the present day rock masses always occupied the same space as they do today. They were far, far greater, were soft and flowing and have gradually collapsed, shrunk and dried up. In the living, widely diffused sphere of the Earth there gradually developed the first lower forms of animal life. We find in the dark roofing-slates crab-like animals, the trilobites, now extinct. We find shells, snail-like creatures and fishes. The Devonian (armoured) fishes which have an outer skeleton are especially characteristic of the shales.

Towards the end of *this* shale period, there appears something which shows that the whole period is subject to an overruling vegetative forming force – the Coal Measures. We have pointed out earlier (Chapter 1) that the formation of coal with its enormous seams cannot be explained merely as the remains of forests. Coal has not arisen out of actually formed plants but out of a prolonged period of shedding and casting off which fundamentally was somewhat *similar to the formation of rock*. In the Coal Measures we have a visible documentation of the transition of the mineral-plant world to the later higher-plant world. Here it is not only rock that is precipitated or cast off from the living processes of the Earth, but organic material which remains as carbon. We find this carbon as graphite and traces of coal in shales older than the Coal Measures. It is this carbon together with traces of iron that often gives to slate its dark colour. In the Coal Measures one finds all the transitional stages from shale to pure coal. One can see from this that there is a real transition from rock-forming to coal-forming. The living process behind it becomes more and more of a purely vegetative nature. It is comparable with the growth of wood in our trees which forms rings year after year. What was already visible as texture in gneiss, crystalline schists and slates now becomes quite evident: *The whole Earth has the tendency to become a tree*. In slate, shale and coal formations we are dealing with a 'tree-stage' of the life of the Earth.

In discussions with teachers at the Waldorf School in Stuttgart,

Rudolf Steiner emphasised that coal does not originate from plants of solid substance such as we find on Earth today. What one finds as leaves, bark, trunks and other plant forms in coal are only emergent transitory forms which, as they arose, immediately *fell away* and became petrified. They are semblances of life which came about like wind patterns in the sand or ripples on the sea shore. Perhaps one might say they are the first attempts in creation to mould elaborate plant forms. What became coal originated in the sphere of life at the outermost periphery of the fluid layer of the Earth. The rhythmic character of coal formation can be recognised by the enormous number of successive seams. Near Aachen there are 45, and in the Donetz basin in Russia as many as 225 seams, one above the other and separated by other strata. The greater the number of seams the thinner they are. The average thickness of coal seams is from 30 to 125 cm, but there are seams as much as ten metres thick.

At the close of the Carboniferous period came a time of mountain upheaval, a period of Earth activity that can be compared to volcanic activity today. We will return later to the question of this early volcanism. Mighty movements of the partly solidified Earth caused the breaking up of whole mountain ranges, the debris later hardening as sandstones and conglomerates. But these phenomena which increased in intensity throughout the slate period and then died down, are closely related to what we have designated as the 'tree-stage' of the Earth. In order to understand this we must take a look at the nature of present day trees.

One can appreciate the special character of a tree if one compares an ordinary annual such as Borage, with its beautiful blue flowers, with a Norway Spruce (our Christmas Tree). Borage is a juicy, hairy herb with a soft, weak stem. It makes a profusion of seeds but the first frost kills it and in the Spring the seeds produce new plants which grow and flower quickly.

In contrast the Norway Spruce takes many years before it flowers, and it needs a further year before the seeds in the cones are ripe. It does not mind the winter, it does not even shed its leaves but remains evergreen. For its existence the spruce is independent of the seasons. Only for growing and flowering does it need the Spring and Summer. This shows that the spruce, like all conifers, must have arisen when there were no seasons as we know them. This means that the Sun and Moon stood then in a different relationship to the Earth. Now, at the time of the origin of the conifers the Earth and Moon were still one cosmic body. It was the time which we have just discussed, for it was during the slate period that the Sun separated from the Earth and this caused the transition that we have described as the Carboniferous period. Of course, the conifers of those days were not the solid forms of today but the first trees gradually began to take shape after this time. They are preserved for us for example in the beautifully coloured, silicified wood and tree trunks of the Permian formation in Arizona, U.S.A. The soft, pulpy wood was impregnated with the fluid silica and completely transformed into agate.

But we have to come back to our present day trees. In one of his lectures to the workmen at the Goetheanum (31 October, 1923) Rudolf Steiner, when speaking of the nature of trees, describes wood as a hardening and drying of what rises as fluid from the roots. The sap draws up minerals which are deposited as dead matter. In his Agricultural Course he compares the tree with a mound of soil on top of which plants grow. One knows from experience that a pile of soil, raised above the level of the surrounding ground, becomes more fertile owing to the better penetration by air, moisture and heat.

Every gardener makes use of this when he builds his compost heaps. Certain life processes are initiated which form humus from plant remains and soil. We have already mentioned this. In the tree – which, according to Rudolf Steiner, we should imagine as a 'mound of soil' this also occurs. In the periphery of the trunk, which is solidified 'fluid soil', bark begins to form, and under the bark the cambium layer. This cambium is a very complicated thing. It arises from the sap which flows down from the leafy crown. In this condition one can recognise it in many trees as gum or resin. If however it remains *in* the tree it becomes cambium which constitutes a sort of amorphous root. From this living layer the wood develops inwards, the skin outwards. This outer skin is however not yet the dead material that we know as bark. What this skin is can best be seen if one goes to a forest in winter and scratches the outer layer of a young sapling which as yet has no dead bark, or even of a mature beech tree which never develops a real, thick bark. Then one discovers that under the skin there is a *green* layer which, even after leaf-fall, still contains chlorophyll and is alive. This is the true skin that envelopes the whole tree trunk like a giant leaf. On the outside of this skin is formed the dead mineralised bark. This consists of deeply wrinkled or furrowed woody material containing tannin and certain oils. What is this bark in reality?

We have seen how, from the cambium and skin, there develops wood to the inside and bark to the outside. The cambium can be compared with the root-nature and the green skin with the leaf. But where is the 'flower' of this living mound of soil – this tree trunk? It is none other than the dying bark. This may appear surprising and hypothetical, but it becomes evident in those barks which develop essential oils, for instance the cinnamon tree. Bark is the result of a process of dying as is the flower of a plant, but this dying process does not lead to scent and colour, instead it leads to denser materials such as tannin, resins and oils. If one examines these substances it transpires that in their chemical composition they closely resemble scents and colours in flowers. The 'dying' of the bark comes about through astral forces working from outside, as they do in the flowering crowns of trees and in ordinary herbaceous plants. Leaves and flowers grow on the twigs of the crown of the living trunk or 'mound of soil' as herbaceous plants grow on the solid ground below. The spherical shape of the crown repeats the rounded surface of the Earth.

What we have here described as the nature of a tree is a true picture of a life process that formerly embraced the whole Earth. There was a time when the Earth was still full of life, when it had a sort of cambium or skin layer out of which developed first the schists and slates and then the Coal Measures. Today we have this in miniature as the formation of wood in a tree trunk with its annual growth rings.

Once this 'earth-tree' had attained a certain size and solidity, other forces from outside penetrated the hardened crust and by tremendous movements the first mountains were drawn up out of the Earth. (Rudolf Steiner remarks that mountains and volcanic upheavals are not caused by pressure from within but by suction from surrounding forces.) It was in this way that the 'bark' was formed on the 'earth-tree'. Today we can still study the same forces forming the 'miniature mountain ranges' on the bark of our forest trees. The secret of the trees is that they have preserved in miniature an old condition of the whole Earth.

Now at last one can understand what Rudolf Steiner said in his Agricultural Course about the significance of schists and slates for the higher plants. For the herbaceous plants, these rocks play the same role as does the wood of the tree for its own blossoms and leaves. This means that the schists and slates of the Earth are the 'tree-trunk' on which the herbaceous plants grow. Alternatively one can say that because once upon a time the Earth was a sort of tree, it became possible for most of our food plants to arise.

On the other hand when we considered the rocks that are associated with the *bark-forming* of this ancient 'earth-tree' (the now familiar series from granite to porphyry) it became evident that they are responsible for the *flowering* of herbaceous plants and trees. The flowering of the mineral-plant world and the life of the world-plant-animal had to be turned to stone in order that from the dead body of the 'earth-tree' the future flowering plant could arise surrounded by bees and butterflies.

6 *The 'Animal-Nature' and Limestone*

No other rock in the world shows its origin so clearly as limestone. Consider for example the Carboniferous Limestone. In the early days of the Earth's history (Pre-Cambrian) one finds limestones, marbles and dolomites which owe their origin to lower animals such as coelenterata (plant-animals) and molluscs, as well as lower forms of plant life such as some algae which we have characterised earlier as transitional forms between plant and animal.

The period in which these first limestones appear corresponds with the beginning of the schist period. All over the Earth one can observe that, wherever gneiss or crystalline schist follows granite or one of its metamorphoses, there are also to be found limestone and dolomite between these other rocks. Naturally they are accompanied by greenstones and porphyries in their varied forms, for this is the time when the great differentiating of life began.

The traces of life in these older limestones are very sparse. Owing to their softness and delicacy, their forms were destroyed. The gelatinous condition of matter also contributed to this destruction. However, there is no need to search for concrete animal forms. The organic origin of these limestones is clear from the great quantities of carbonic acid which they contain. Another substance of organic origin that plays a great part in these old limestones is graphite. Although as pure carbon it points more to a vegetative process, one must remember that it was just at this time that the plant-animal began to differentiate into the plant and the animal. There was as yet no clear division between the various life processes and forms, and therefore there is a confusing interpenetration of rocks. The enormous scale of those processes and formations can be seen in the province of Ontario, Canada. There we have a region where these crystalline limestones and marbles have a thickness of 15,000 metres. Between the limestones and the often coloured marbles there are layers of quartz, gneiss and serpentine. The marble itself contains mica, hornblende, serpentine and above all, graphite. The graphite content of the limestone may be between 3% and 10%. There are seams of graphite of from one to four metres thickness. *In this area there appears to be more carbon (as graphite) than was deposited (much later) during the whole Carboniferous period over the entire Earth.* If one could visualise how much coal is extracted from the Earth daily and how much more is still lying in the Earth, it might be possible to form an approximate picture of the superabundance of life forces behind the forming of these limestone and graphite masses.

What meets us in these early times as limestone without fossils is the first sign of a mighty process that drew the animality down to the Earth with the help of calcium. Rudolf Steiner goes into this process in his lectures *Mystery Knowledge and Mystery Centres* which we have mentioned earlier. One must not imagine that this process only began when the 'plant-nature' emerged. The life history of the Earth is not a succession of events that can be followed step by step in the layers of rock like turning the pages of a book. It is more of a mingling, an interpenetration of events. It is however quite noticeable that certain life processes clearly predominated at certain times.

From these early times onward, limestone is on the increase while silica rocks decrease. *This means that the animal emerges and makes itself felt.* From early periods onwards through the Cambrian, Silurian and Devonian periods (which are the real slate/shale periods) up to the Coal Measures, we find, between the slates and sandstones, increasing layers of limestone and calcareous rocks. When the 'plant-nature' reached a certain climax in the forming of rocks, calcium took the upper hand. This was after the Bunter Sandstone.* In the Muschelkalk we encounter solid limestone masses which extend over wide areas like massive underground masonry. Where Muschelkalk is worked in large quarries we can study its remarkable structure. It consists of bands of dark bluish limestone of thicknesses up to about half a metre, alternating with narrow, darker bands of marl a few centimetres thick. Vertical jointing gives the impression of masonry. The colouration is due to very finely disseminated iron sulphide and bituminous substances. Between these beds occur

*See footnote Chapter 9 p. 82.

thick bands of harder dolomite (calcium magnesium carbonate). This dolomite is often silicified. In some layers one finds banks of algal concretions (produced by blue-green algae) and in others quantities of mussels, terebratulids (ancestors of the lampshells which today live only in the depths of the sea), and sea-lily stem sections. These visible animal remains of the Muschelkalk are very poor in species but very rich in individuals. The name Muschelkalk originates from the fact that these few species, especially the mussels, are present as fossils in huge quantities.

The most remarkable feature of the Muschelkalk is its alternating layers of limestone and marl which make it look like solid masonry. When one realises that the Muschekalk area in Europe originally stretched from Heligoland to North Africa and from western Spain to the Caspian Sea, one wonders how it was possible that this rhythmic stratification could take place over so large a surface. The thickness of the formation varies from about 10 metres at the edge of the area to about 250 metres in the middle!

The stratification of the Muschelkalk resembles somewhat that of the schists and shales. But instead of siliceous formations we have here mainly calcarious rocks alternating with the layers of marl. (Marl is more or less calcareous clay). The calcium in these strata is undoubtedly of animal origin and the bands of marl contain much iron sulphide and bituminous substances which account for the dark colour.

This rhythmic stratification is a real puzzle for modern geology and one that cannot be solved by the application of present day physical and chemical laws.

In reply to questions put to him by the workmen at the Goetheanum Rudolf Steiner gave a lecture (23.9.22) *Earlier Conditions of the Earth.* In this lecture he spoke of the time which immediately followed the event of the separation of the Moon from the Earth. There was no solid rock on the Earth at that time. Everything was there in a fluid and gelatinous state in thick layers. Later it shrank, dried out and solidified as rock. Rudolf Steiner described gigantic forms of life which lived in this viscous earth. These gigantic creations which were so big that one could have superimposed the whole of France or Switzerland on their backs were somewhat oyster-like with a scaly surface resembling the shell of a tortoise. These great animals secreted these shells on their upper surfaces, and left behind a slime as snails do today. However they left no fossil remains.

What is important for us in this picture is the idea of a *life process* that led to the formation of the armour-plated shells of these animals. This process was not localised anywhere in particular but extended over the whole Earth and was part of the life of the Earth.

This forming of a carapace like that of tortoises and oysters is a *life process* of the Earth as a whole, as was the forming of schists. Just as we cannot consider schist as the *remains* of trees so we should not consider the regular layering of the Muschelkalk as the *remains* of giant 'oysters'. Instead, the remarkable structure of the Muschelkalk with its alternating bands of limestone and marl becomes intelligible as an *organic* process retaining something like an after-image of the carapaces of these giant animals.

If one studies a present-day oyster shell one finds that it is built up exactly like Muschelkalk on a small scale. In the oystershell, layers of lime alternate with a remarkable substance called conchioline. It is related to the chitin of the exoskeleton of insects and is composed of albumin and cellulose. Every thin flake of lime in the oyster or snail shell is enclosed in a thin veil of this conchioline. The layers are arranged like roofing tiles overlapping each other. The construction of the Muschelkalk is similar except that everything is on a giant scale and through petrification the organic origin of the stratification is obscured.

The remarkable stratification of the Muschelkalk is a 'picture' of life processes into which inorganic matter subsequently penetrated. It is fundamentally the same phenomenon as in moss agate which we described in the chapter on the plant-animal and its signature. There as here, mineral matter seeped in, leaving a picture of a former life process in the discarded sheath.

What we find in the Muschelkalk as actual animal forms – mussels, snails, crinoids, etc. – stems from a later time and lived in the discarded remains of the ancient life.

Thus in the Muschelkalk we have found a picture that represents a life process that we might call the 'oyster' stage of the Earth. It was a preparatory stage for the further differentiation of the animal world. The giant forms disappeared and the life of the Earth as a whole dispersed itself into tiny forms which subsequently built up the mighty calcareous layers of the Jurassic and Chalk periods. At the same time there gradually arose the new giant forms of the Saurians. Here lies the beginning of the new transition that leads to the mammal era. After the precipitation of calcium in the Chalk, the 'life' of the Earth begins to fade away. The rocks of the Tertiary period (Atlantis) can only to a very limited extent be considered as new formations of the living Earth. From the middle of this period dissolution set in; the breaking up of what had been formed.

*

In the foregoing we have endeavoured to sketch the life processes connected with calcium carbonate. It should be noted that this is the form of calcium which is associated with the lower animals and the lower plants (algae). These lower organisms are the first forms of life that have been preserved even in the very oldest strata.

In the lectures on *Mystery Knowledge and Mystery Centres* previously mentioned, Rudolf Steiner gives a very vivid description of how these first animal forms arose in the soft fluidity of the Earth. In the rarefied, watery albuminous atmosphere, previously described, existed not only vegetative processes, but also the first attempts at animal forms, as transitory and fleeting as the plant forms. He speaks of how the lime had the capacity to evaporate like water after being incorporated in the fluid Earth. There was a rhythmic up and down movement of lime as vapour and rain. As the lime descended and densified it enwrapped the soft animal forms and drew them down to earth.

Thus these first tender, soft, 'fluidic' forms were clothed with shells and protective armour and appeared on Earth in more solid forms. As molluscs and bony-plated fish we find them in the old strata. Many molluscs remain little changed to this day but the bony-plated fish have died out. They were an early form of a group which has since developed further.

Though Rudolf Steiner's description of the evaporating and precipitating lime may appear surprising, yet lime still has this property today. We have to remember that this process did not take place in an atmosphere such as ours, but in a 'fluid' atmosphere. Naturally lime can no longer clothe fleeting animal forms and bring them down to the earth, but it still has the peculiar capacity, in lower temperatures with the help of the carbonic acid of the air, to dissolve in water and become bicarbonate of calcium and, when this water is heated, to precipitate out as solid lime. This is a process continually taking place in nature which plays an extremely important role in the maintenance of life in plants and animals. Imagine, if lime were as insoluble as silica, no living organism could take up this substance which is so important both for support and nutrition.

There is another form of calcium in rocks which quantitatively cannot compare with calcium carbonate. It is calcium phosphate. Whereas calcium carbonate is mainly the result of the living processes of lower organisms, calcium phosphate is the skeletal substance of vertebrates and of man.

In rocks calcium phosphate appears in two forms. First as the well known apatite (natural calcium phosphate plus either fluorine or chlorine) and as phosphorite. Apatite is a component part of pegmatite and many of the greenstones. One can also find it in the younger basalt and other lavas. On the other hand phosphorite has formed in sedimentary rocks (sandstones, shales, marl and limestones) from bones and proteins of dead animals. It is unmistakably of organic origin.

The mineral apatite, which often appears in wonderful crystals in the granular igneous rocks, has exactly the same chemical composition as the substance in animal and human bone. This is a very remarkable fact, for here the higher animals and man have incorporated in their bodies a very clearly-defined *mineral* process. One can recognise the crystalline structure of apatite in a thin section of bone. It is present quite independently of the bone cells.

Now we know of two metals in particular which are associated with apatite – iron and tin. The enormous deposits of magnetic iron ore found in northern Europe, Lapland and northern Canada are completely permeated with large quantities of apatite. And in this apatite are to be found minute quantities of lead phosphate.

If one looks at human bones the way Rudolf Steiner has done in his medical lectures, one finds that these three metals – iron, tin and lead – play an important part in the forming and functioning of the bone system.

From the interior of the bone iron finds its way into the newly created blood. In the joints which bring movement into the rigid skeleton – and which are filled with joint fluid and thus maintain a balance between fluid and solid – tin is active. (This does not mean that tin is present as a substance). In the hardening of the bone, lead is active. Chronic lead poisoning produces calcification of the blood vessels and, in certain circumstances, an excessive hardening of the bones. In such cases lead is deposited in the bones. However in healthy, normal bones one does not find lead.

In apatite we have another example of how an organic process – in this case bone formation and function – is the prototype for phenomena in the apparently lifeless mineral kingdom which otherwise we would find strange and puzzling.

7 *The Human Being and Salt*

Up to this point we have surveyed in broad lines a major part of the rocks of our Earth and have tried to show how these now dead formations have in the past emerged from life processes of the *whole Earth.* We found that these once all-embracing processes appear today transformed as certain phenomena and forms of life in the various kingdoms of nature.

When Rudolf Steiner links with the forming of salt that being who has cast off and left behind the animals, plant-animals, plants and minerals – namely man – it is not to easy to see the connection as with the other rock formations. If one wishes to understand this connection, one must turn ones attention to the significance of salt in the life processes of man. Not only what contributes to his nutrition, growth and regenerative powers, but that other part of his life that we know as consciousness, imagination and thinking.

In order to appreciate what far-reaching significance salt has for man's whole constitution, one has only to deprive a healthy person completely of the salt he normally uses for seasoning. The first symptom that appears is loss of appetite. This is due to the fact that salt has the peculiarity of bringing out the individual tastes of all foods and, through the experience of taste in the area of mouth and tongue, it works deeply into the unconscious functions of the inner glands connected with the stomach and intestinal digestion. Thus, through salt, something becomes conscious that works unconsciously in digestion and nutrition. The *inner* taste is stimulated which, according to Rudolf Steiner, extends right into the liver. This *inner* taste, which only appears consciously as appetite, is linked with another peculiarity of salt. It guides the individual nutritive substances to the appropriate places in the organism.

The reduced fluid requirement of a person on a salt-less diet shows that the organism needs to retain the salt present in the body fluids. Normally the salt in the blood and tissue fluids would be continually exchanged. In this process however, which regularly introduces new salt into the system, lies one of the most important biological functions connected with digestion and nutrition.

The withdrawal of salt has another significant effect on man. He becomes tired and apathetic and finds that his ability to think is impaired. These phenomena are all connected with the other side of his life – his consciousness. The nature of these two processes is diametrically opposite. Digestion and nutrition build up, whilst consciousness arises as a result of a breaking-down process. Physiologically, consciousness and self-awareness destroy what is built up and this breaking-down takes place almost entirely in our nerves and brain. In this nerve-sense system, as it is described by Rudolf Steiner, man is able to concentrate nutritional substances to the point of mineral density and lifelessness, but he is also able to dissolve again, destroy and cast them out from his organism. These physiological 'death-processes' are specifically human, they do not occur in animals.

In olden days this was called the 'salt-process'. It signified that out of something in solution a hard substance was precipitated. In this sense one could even call the forming of all rocks and minerals of the Earth a 'salt-process'. But one must note that most rocks and minerals are *not* soluble in water – at any rate not to the extent of real salts. Only true salts are soluble in water.

Thus in man we find salt in two conditions; first in solution in the blood, and then as *process* in the brain and nerve system which leads to a sort of depositing, and which is the physiological basis of consciousness. The most well known example of this is the brain-sand in the pineal gland which, if it is absent or abnormal, causes idiocy or feeblemindedness.

Where salt is present in solution as in the blood and digestive juices of man and animal, we have the primal condition of salt as we find it in nature in sea water. The oceans of the world are but the remains of that albuminous atmosphere of the Earth which we have described earlier. In this 'amniotic fluid' of the living Earth the living forms of the kingdoms of nature developed. They cast out the mineral part into the mineral world and drew into their systems a part of this living atmosphere as blood and the other body fluids. In this way was the original life of the Earth itself divided up into individual forms in the kingdoms of nature. When the old albuminous atmosphere gradually decomposed, there arose out of it the present-day waters of the oceans and the atmosphere of air. When albumen (protein) decomposes there appear the substances present in seawater and in air, namely: salts, water, oxygen, carbon dioxide and nitrogen.

The oceans are still today a real 'amniotic fluid' for a multitude of organisms. It is a 'physiological fluid' in which the oldest and most primitive forms of life could persist till the present day. How near to life the substance of sea water is can be seen from the proportions of soluble salts in it – especially sodium, potash, magnesium and calcium salts. These proportions resemble those of human and animal blood. One can therefore use pure sea water, diluted to the salt level of human blood, to mix with medicinal extracts of plant or animal origin for injection into the blood stream.

In contrast with the primal condition of salt, even today closely connected with life, we must also consider quite a different process which has deposited salt in the Earth in solid crystalline layers like rock.

The enormous beds of rock salt in the interior of the Earth are distributed over all continents and are found from early Cambrian and Silurian times up to the Tertiary system.

The beginning of this deposition of salt in the Earth coincides more or less with the differentiating of living things and of rocks that we spoke of in connection with slates. It is the 'time' when limestone and dolomite began to appear. The deposition continued through the subsequent formations reaching a first *maximum* immediately after the Coal Measures in the Permian system. The famous potash and rock salt beds of Stassfurt belong to this system.

A further maximum in salt formation occurred in the Tertiary system when the forming of *new* rocks and strata from original life processes was nearing its end. The potash beds of the Upper Rhine basin, Poland and Spain belong to this period. When considering the great quantity and widespread occurence of rock salt one has to bear in mind that salt is a water-soluble substance which has taken on the appearance of rock. But we must also point out that though salt today is soluble, it does not mean that these huge beds were formed from a fluid solution such as sea water. Rudolf Steiner makes the significant remark that this solubility of minerals which we observe in salt is, *in the mineral kingdom, the most recent characteristic that has developed*.

This observation becomes intelligible when we bear in mind that all rock was originally in a gelatinous colloidal condition. Gels and colloids are original states of matter, *not solutions*. They have a close kinship with living matter – protein.

The whole appearance of the salt deposits suggests that this 'salt-rock', like the slates and coal measures, has been cast off from a major life process of the Earth. Rock salt occurs in layers and – as for instance in the Permian (Zechstein) salts – frequently alternates with potash and magnesium salts interspersed with thinner or thicker layers of anhydrous gypsum. The salt deposits are almost everywhere encased by massive layers of anhydrites which are, as often as not, accompanied by dolomite. These enveloping rocks are impermiable so that neither surface nor underground water can normally penetrate to the salt beds. So we have the curious fact that a soluble substance in the interior of the Earth is thus protected from contact with water.

One can imagine what it would mean if all the salt buried in the land masses were exposed to the influence of water, since the quantity of rock salt equals approximately the amount of salt in solution in the oceans – thousands of billions of tons. Since about 70% of the Earth's surface is ocean and barely 30% is land, the rock salt, if it were not protected, would impregnate the soil with a salt content rendering all life impossible.*

The fact that this does not occur points to a rational organic process in the forming of salt beds. This process is connected with man in the same way as is slate with the plant, and limestone with the animal.

In Rudolf Steiner's notes mentioned in Chapter 2 he says 'in salt the universal human being is extinguished'. This means that the first appearance of salt is connected with the fact that the universal being 'man', until then part of the Earth as a whole, now begins to differentiate into individual human beings. In order that the individual man could develop further out of the 'World-Man-

*The salt and mineral content of the oceans alone would cover the whole Earth with a layer 45 metres thick.

Being', the animal nature, the plant nature and the plant-animal had to have 'died away' and the 'mineral be burnt up in sulphur'. At the same time there began the differentiating of the kingdoms of nature which is reflected so strikingly in the diversity of the rocks. Together with porphyry, schists, slates and limestone, salt also appeared early as an expression of the fact that now a being began to act that had the capacity to overcome the hardening of matter through the dissolving power of living water. Man is the only being able to make use physiologically of the mineralising decomposition in the nerve and brain system in order to develop consciousness of selfhood and power of thought. The reflection of this human capacity in the world outside man is the ordering of salt deposits among rocks. Before man completely descended to the Earth he 'condensed thoughts in the salt.' Through his power of thought he banished the death forces of salt into the Earth. This salt was in a condition that had nothing to do with its present solubility. It was still a colloid that obeyed organic laws.

The concept that a substance which today is highly soluble in water was once a gluey colloid that swelled as it absorbed water but did not dissolve, is certainly very unusual. There are however present day phenomena that vindicate this idea.

In many salt deposits, especially in rock salt, we find, in between layers of coarse salt, lenses or layers some metres thick of 'clear salt' of nearly glass-like transparency, hardly streaked or clouded at all. If one puts a light behind such a block in a mine, the whole block is illuminated like a vitrious mass and one can see the source of light through the clear salt. If the salt had crystallised out in lagoons and shallow bays under the influence of tropical desert temperatures – as postulated in the great inundation theory of Oxenius and van t'Hoff – it could never have this clear structure, but would be deposited as a crumbly mass of small crystals.

There is another phenomenon which points to the malleable-colloidal condition of salt. In 1937 when prospecting for oil was going on in the Hanover-Celle region, a boring was made which entered a salt dome at a depth of 475 metres and penetrated to 3818 metres without reaching its bottom. At the depth of 3818 metres the salt had a temperature of 130°C and was a dough-like malleable mass in which the drill got stuck. Since the melting point of rock salt is over 800°C this is a strange phenomenon which sheds significant light on the state of salt in the depth of the Earth. In this connection it is interesting that, when making the Simplon Tunnel, crystalline quartz was found in the heart of the mountain together with a gelatinous colloidal silicic acid – a silica that was still in a similar primal state (before crystallisation).

With the finding of the above mentioned salt dome over three kilometres thick we are confronted with the tremendous size of these salt beds. Hundreds of such salt borings were made in northern Germany and one now knows that the Zechstein salt beds alone cover an area of about 100,000 sq. kilometres. They extend from the coast of the North Sea between Bremen and Lübeck, eastward to Berlin and south to Erfurt. Potash and magnesium are extracted from about 250 pits and converted into fertilisers and chemicals.

These Zechstein rock salt beds – only the upper horizon is accompanied by the valuable potash salts – lie at depths of some hundreds to over a thousand metres. Below them usually lie still older layers of rock salt which however are not worked since there is enough pure salt elsewhere more easily accessible.

In Europe the Zechstein salt beds stretch from this German area westward to England and eastward to Poland, while 2000 kilometres further east lie the great salt beds of the Urals.

The Zechstein salts constitute only the older salt beds of this area. At higher horizons we have the salts of the Bunter Sandstone and Muschelkalk. These last extend from the district of Magdeburg through Thuringia, Schweinfurt, Heilbronn and the eastern Black Forest to the Swiss border.

Very young tertiary postash and rock salts occur in the Upper Rhine basin between Basle and Freiburg at a depth of from 500 to 1000 metres. Enormous deposits of salt are found in Russia especially at Kazakhstan where 1500 to 1800 individual occurrences lie only a few hundred metres below the surface but extend to depths of 5000 metres. The Dossor one is 12 kms. long, 8 kms. wide and 5 kms. deep. The Iskine one occupies an area of 30 sq. kms. and is one and a half kms. thick in the south and about 5 kms. in the north. Another near Osinki covers 70 sq. kms.

Through Russian Central Asia there are thousands upon thousands of larger and smaller salt occurrences. They link up with the salt beds of Turkey, Syria, Irak and Palestine. The last includes the Dead Sea with its forty thousand million tons of salt in solution, one of the strangest and most remarkable salt phenomena of the Earth. One of the richest areas is Iran and the islands of the Persian Gulf. On these there are open cast salt mines or rather mountains 5 to 10 kms. thick with furrows of salt glaciers plowing through them. India is famous for the Salt Range east of the Indus where very old salt lies in enormous quantities to a great depth.

The great salt beds of North America are to be found especially in New Mexico and Texas where deep borings for oil revealed beds of salt 300 to 500 metres thick over an area of 200,000 sq. kms.

.In the course of prospecting for oil in the Gulf of Mexico hundreds of salt domes were encountered which were so thick that drilling was abandoned. The 'Five Islands' salt masses are famous. They tower over the plain and extend underground to a depth of several hundred metres.

We have mentioned only the most important salt formations which can perhaps give some idea of the tremendous and remarkable phenomena of the salt rocks of the Earth.

8 *The Mystery of Oil*

Having dealt with the mineral-plant and plant-animal stages we must now intercalate something else that is inseparable from the Earth's life – the origin of oil and its related substances such as asphalt, bitumin and mineral wax.

Numerous theories have been put forward to explain the appearance of these substances in enormous quantities in certain regions and during certain geological periods. One of the first theories attempted to attribute the origin of oil to purely inorganic processes such as can be produced in the laboratory from metal compounds of carbon i.e. carbides which, in contact with water, become hydrocarbons. Such hydrocarbons are for instance methane (marsh gas) or acetylene, which arises when water causes the decomposition of calcium carbide. But since oil consists of far more complicated compounds, this theory was soon dropped. Besides, in the meantime, it had been shown that the formation of complicated compounds from these simple hydrocarbons could only take place at high temperatures and under high pressure, and high temperatures were out of the question.

Later it was assumed that oil was formed by a sort of distillation of organic remains such as coal and peat through volcanic action in the interior of the Earth. This theory too had to be discarded when it became ever clearer as new oil fields were discovered that oil is never found far from the place of its origin and no volcanic rocks are ever present in the vicinity. It is a peculiarity of oil that it does not appear until a certain period and then exclusively in sedimentary rocks such as sandstones, sands, shales, marls and limestones. It has never been found in metamorphic rocks which are mostly distinctly crystalline. At the present time it is generally accepted that oil is of organic origin. It can be traced back to a transformation of plant and animal remains whose proteins and fats, under the influence of pressure and mild heat throughout long periods of time, became present-day oil. To these factors the latest theories add the activity of certain bacteria which have been found in the salty water of oil wells.

These oil bacteria have definite pecularities. It has been known since the end of the last century that there are bacterial organisms which can create simple hydrocarbons such as marsh gas out of organic remains under anaerobic conditions. The metabolism of these anaerobic bacteria does not need the free oxygen of the atmosphere because it is able to take this vital substance out of organic remains such as lignin, protein or fats (which all contain oxygen bound up in some form). The organic matter is thus broken down into hydrocarbons and water. In contrast, aerobic bacteria need atmospheric oxygen to live, and give off carbon dioxide. Our yeasts which 'ferment' sugar into alcohol, carbon dioxide and water, work in a similar way.

During the first decades of this century bacteria were discovered which could 'digest' rubber and some volatile hydrocarbons such as benzol, toluol (phenol methane), etc. It was then noticed that certain anaerobic bacteria remained alive in hydrocarbons and did not die off. In the middle of the twenties bacteria were first found in the oil wells and their salty water. These bacteria came from a depth of many thousand metres and must have existed in the interior of the Earth for countless millenia. They have remained alive in the salt solutions of varying concentrations which accompany the oil. The high pressures presumed to prevail at these depths had not been able to destroy them, nor had the relatively high temperatures.

Under laboratory conditions it was established that in an artificial nutrient, some of these bacteria were able to multiply entirely anaerobically at temperatures below freezing point and others at temperatures of 85°C.

Many of these bacteria were completely new varieties quite unknown on the surface of the Earth. Some strains were able to break down any organic material into petroleum substances. Others again took these and metabolised them into other simple hydrocarbons. In effect researchers were faced by a plethora of new phenomena for which no common denominator could be found. Furthermore substances were discovered in oil that inhibited the growth of bacteria but did not kill them. Closer examination of these inhibiting substances revealed the surprising fact that they were organo-metallic compounds containing heavy metals such as copper, nickel, iron, molybdenum and vanadium. These compounds closely resemble haemoglobin, chlorophyll and the curious substances found in the blood of molluscs and echinoidea, (haemocyanin which contains copper) and the blood pigment of the sea-urchins (containing vanadium). Also there is a red colouring matter containing molybdenum in the nitrogen bacteria attached as nodules to the roots of leguminous plants which is very similar to the haemoglobin in the blood of higher animals and man, which however contains iron.

In the early forties when all these phenomena were discovered, a group of American scientists followed up the problem to see where it would lead. During this research the question arose as to why, in the course of millenia, all oil had not been 'devoured' by these oil-digesting bacteria and transformed into simple hydrocarbons such as methane, etc. They tried to explain this by the presence of salt water, the above mentioned inhibiting substances and the oil itself.

Scientists consider that the answer to this question has not yet been found. But in fact the answer is provided if we follow what happens when certain methods are employed in boring for oil.

When sinking a well it is necessary that the drill hole be continually flushed with water so that the fragments of rock loosened by the drill are washed away. When the first traces of oil are found in the water brought to the surface, one knows that one has 'struck oil'. In order to wash away the debris more effectively certain substances have often been added to the water. It has been observed in certain cases in the denser rocks, that after an initial flow of oil it gradually dried up again. By pumping water into the bore hole to force the oil out of the porous rock the opposite was achieved: the oil well dried up in spite of the fact that great quantities of oil were present.

After much investigation it was found that the water reawakened the bacterial world to new life and that the products of their metabolism so changed the viscosity of the oil and salt water that the pores of the rock were blocked and the outflow of oil prevented.

This fact is of far-reaching importance for the understanding of bacteria and their activities. It shows that bacteria in oil and its associated waters are now in a state of arrested activity, but this does not mean that they were always inactive. The tremendous quantities of gas present in most oil fields and consisting mainly of methane show quite clearly that bacteria indeed have transformed part of the oil. But this past life process in the depth of the Earth has come to a halt – to a certain equilibrium. The moment conditions are altered for the quiescent bacteria – by pumping in water – their activity is reawakened.

Thus we find in association with oil an extraordinary rich and varied bacterial life which is apparently in a more or less quiescent state. In oil itself one finds substances which are closely related to certain vital substances familiar to us in nature today. Apart from the organo-metallic compounds, there are also some that have the effect of oestrogenes i.e. they act on the sexual organs of the higher organisms as do hormones. The remaining constituents of oil are compounds which are reminiscent of resins, waxes and essential oils of plants, together with substances containing sulphur and nitrogen more suggestive of animal proteins. If humus-like materials are present, the oils are dark and rich in asphalt (bitumen).

If one examines oil in its crude state, or isolates single constituents, one observes in each case that they have optical properties. A polarised ray of light is rotated when passing through the fluid. This is a pecularity of substances which originate from life processes. Inorganic and synthetic ones do not show it.

Since it is now certain that oil has arisen through life processes, we may ask what sort of a life was it? Was it plant-like or was it more animal? The constituent parts of oil suggest sometimes one sometimes the other. Perhaps we can get a clearer idea if we can picture, even if only vaguely, the enormous extent of this life from which oil originated.

During the 85 years between 1870 and 1955 two thousand million tonnes of oil were extracted from the whole Earth. One cannot visualise such quantities. By careful investigation all over the Earth one knows that more then 10 times this quantity remains in the ground – about 25,800 million tonnes. These figures only refer to oil sources which flow out naturally or can be pumped out because they lie in sands, sandstones or porous limestone. Besides these there are the vast strata of oil-bearing shales which in the U.S.A. alone are estimated to contain 10,000 million tonnes. These oil-bearing shales occur all over the Earth but their oil can only be extracted by mining operations since the fine-grained shales hold the oil like a sponge. The quantity of oil contained in these shales is at least as great as the estimated quantities in sand and sandstones.

Apart from all this, there are thousands of oil wells which, besides oil, produce thousands of millions of cubic metres of gas, as well as borings which yield gas only. All these owe their existence to the activity of the oil bacteria. To add to these inconceivable quantities, there is also the thickness of the oil-bearing strata to take into account. They range from 500 to 4000 metres.

The deepest borehole in the world, in Plaquemines County Louisiana (U.S.A.), struck oil at 6880 metres in 1956. This oil is in relatively young strata. In Kern County, California, there is a borehole which struck oil at a depth of 5457 metres.

Such thicknesses are comparable to those of the old rocks, slates and chalk. If one looks at the geological periods when the oil-bearing strata originated two main epochs can be distinguished. The first extends from the 'slate' period of the Silurian and Devonian systems to the Coal Measures. Its oil content is estimated at about one third of the world total. The remaining two thirds belong to a second epoch which begins round about the Chalk times and extends far into the Tertiary.

Most of the oil of the first epoch lies in the Middle West of the North American continent while the young Tertiary oils are mainly found in the Eurasian continent with the heaviest concentration in the Persian Gulf. Part of the younger oils lie along a line stretching from Wyoming and California across the Gulf of Mexico to Venezuela.

No remains of larger animals or plants are found in the oil-bearing sands, sandstones, shales and limestones. The sands show no trace of petrified fossils such as mussels etc. The limestones are also relatively poor in fossils. Only the shales are often built up in their entirety of minute shells of marine or fresh water organisms such as our present day diatoms or radiolaria.

If oil originated only from these masses of the remains of dead marine organisms, one should find them in *all* oil-bearing rocks.

Conversely there are many deposits of similar remains which contain practically no trace of oil.

Because of this, researchers tried to account for oil by the dying of masses of algae, but there are too many substances of animal origin in oil for this hypothesis to hold good. It may nevertheless be true that the dying of masses of minute organisms such as algae, diatoms or radiolaria is partly responsible for the formation of oil. The major part of the oils of both epochs originated in the 'flooding life' which literally bloomed in the warm watery element of the life of the whole Earth as it was then. Let us think back to the flower nature of the mineral-plant and plant-animal of the young Earth which we described in previous chapters. In oil we have the organic residues of the tremendous disintegration of a life not yet differentiated into individual animal and plant forms. This life followed its course in atmospheric conditions which today we only find in a densified form in the rivers and oceans. It can happen today that in warm waters a 'flowering' as one might call it, takes place. Suddenly, through certain climatic conditions – warm currents, etc. – a tremendous proliferation of lower organisms occurs which rises like a flood for a short time and then quickly dies again. Today oil no longer appears in any quantity from such 'flowering flooding life'. Though in tropical regions one sometimes sees a slight film of an oily nature on the surface of water which shimmers with iridescence, the real production of oil belongs to epochs which had far more exuberance of life than is known anywhere on Earth today.

In the lectures on Mystery Centres previously quoted Rudolf Steiner divides the creation of life into three great periods. From the first no traces of life come down to us, only the oldest rocks of the Earth. From the second we have the old oils (lying mainly in America), coal and the rocks of the slate period (Palaeozoic) with their traces of plant and animal life. From the third period of creation come the younger oils extending from the Cretaceous to the Tertiary systems, the sandstones and limestones of the Earth's 'middle age', and brown coal. In the Cretaceous epoch we find the Saurians, and in the Tertiary the remains of the ancestors of our present day plants and animals.

When, during the Tertiary period – before the beginning of the

Plate 1

Folded gneiss Finstergrund, Black Forest

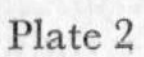

Plate 2

Orthoclase porphyry (large felspar crystals in more or less fine-grained ground mass). Finstergrund, Black Forest

Plate 3

Orbicular diorite (cut at right angles). Santa Lucia di Tallano, Corsica

Plate 4

Orbicular diorite (cut and polished). Kirchspiel, Pöytyö, Finland

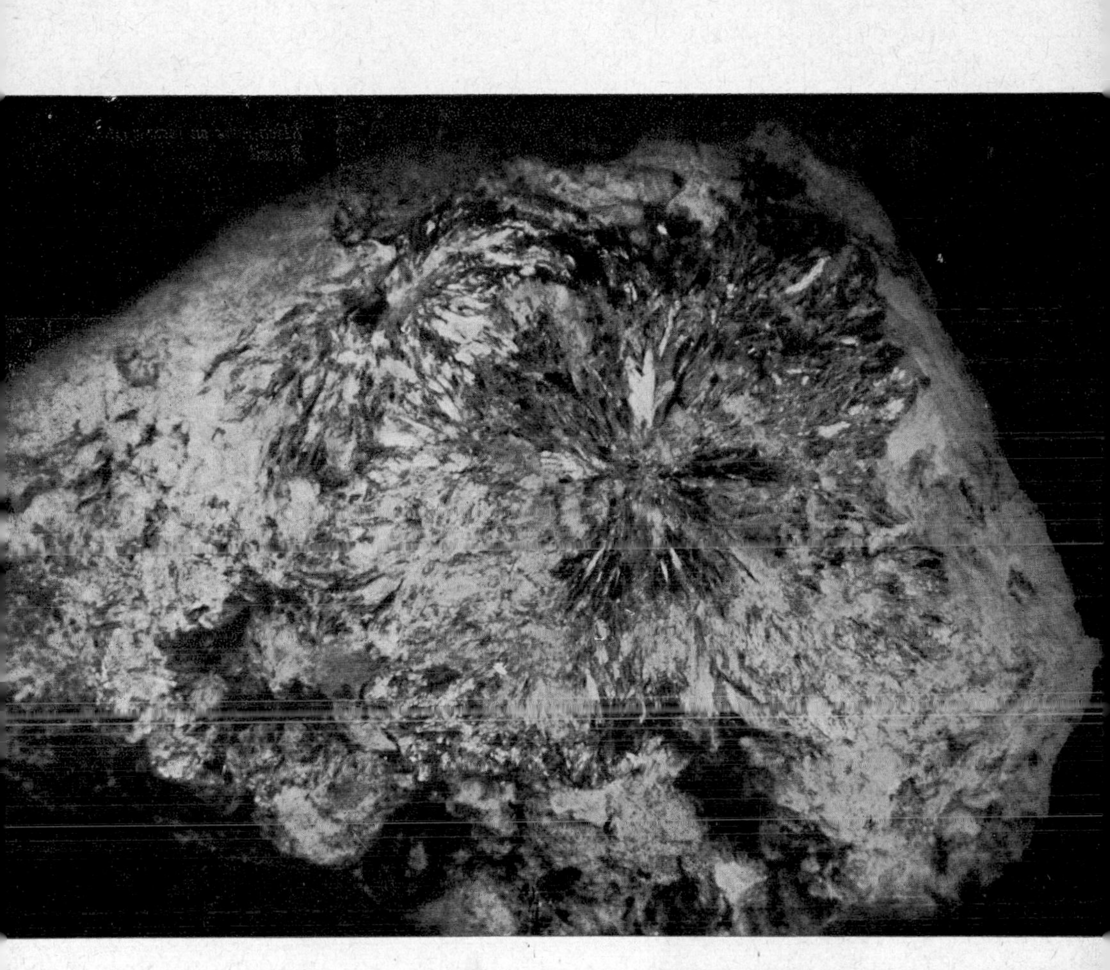

Plate 5

Mica rose (lithium mica) from pegmatite vein in granite. Teufelsküche near Schenkenzell, Black Forest

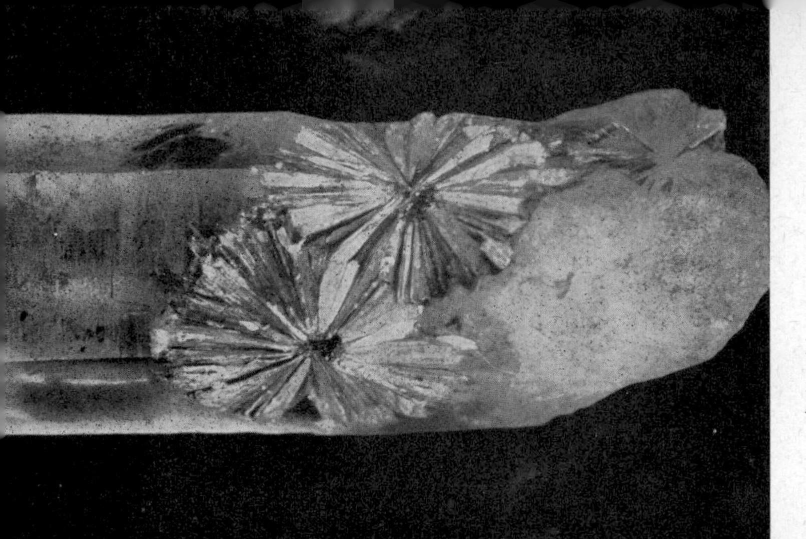

Plate 6

Mica roses on rock crystal. Brazil

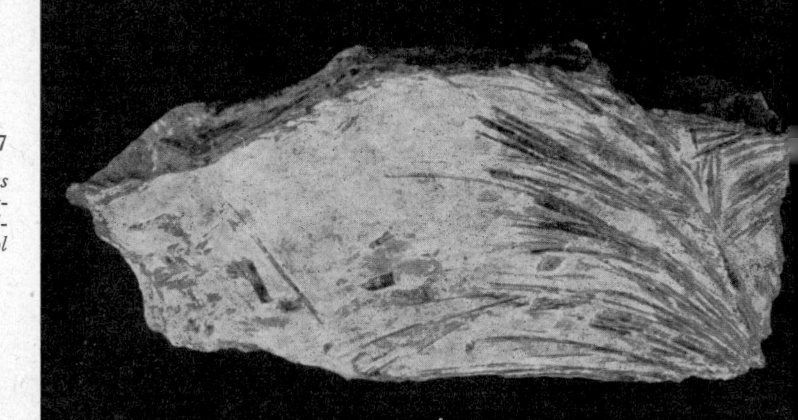

Plate 7

Diopside crystals in crevices of chlorite schist. Schlegeisgrund, Grosser Greiner, Zillertal, Austrian Tyrol

Plate 8

Hornblende garnet gneiss with fans of hornblende crystals. Gotthard Massif, Airolo, Ticcino, Switzerland

Plate 9

Chrysotile asbestos. Canada

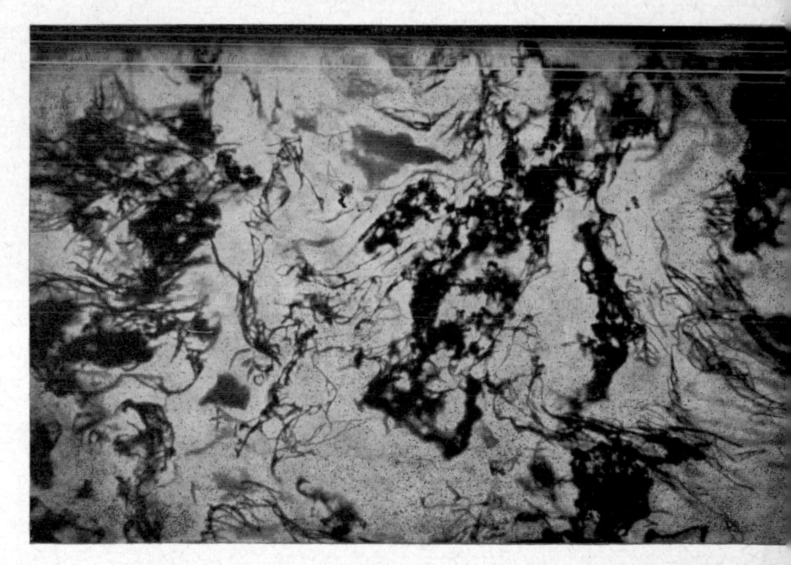

Plate 10

Moss agate, thin transparent slice. India

Plate 11

Meteoric iron, cut and etched. Octahedral crystalline structure (Widmannstetter figures). Fallen 1836 in Bethany, Great Namaqualand, South Africa (natural size)

Plate 13

Basalt columns, Island of Staffa, Scotland. Height of columns 18 metres

The arrows indicate where the small pebbles which caused the impressions are still embedded. ↓

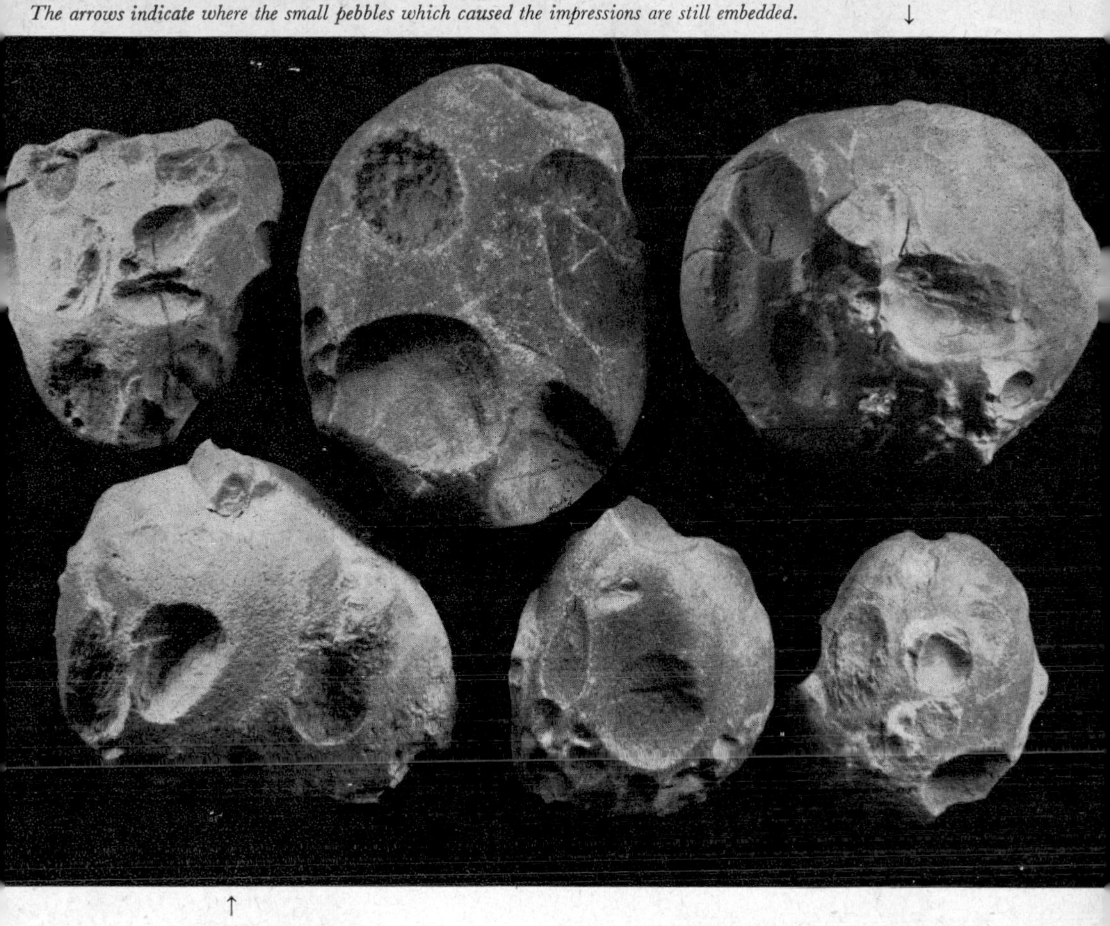

↑

Plate 12 *Pebbles with impressions from Miocene Torton Flysch. Rechtis über Kempten, Allgäu*

Plate 14

Basalt, columnar cleavage.
Rossberg near Darmstadt. (*photo: Hans Walter*)

Atlantean development – the 'life of the Earth itself' began to dissipate, the remaining 'life of the atmosphere' was precipitated in tremendous organic deposits. These are the inconceivably large quantities of the younger Tertiary oils. From their composition they show that they have more of a plant origin. The transformation of these remains of life into oil is still therefore a life process of the *Earth* as a whole. It is a process that we can still see today when our plants produce essential oils and resins. Substances are then created which have a close resemblance to oil. It is therefore not surprising that in one of his medical lectures Rudolf Steiner says that the Earth was able to create oil and that these oils are somewhat similar to plant oils.

In one of his lectures to the workmen (9.9.24) Rudolf Steiner spoke of another kind of oil production. It takes place in the formation of humus and is of importance for the flowering process of plants. We see therefore that this 'flowering flood of life' is still present today though very diminished and transformed. It no longer produces oil as formerly but serves those parts of the plant which are able to produce oils. The oily humus substance mentioned above is also found in the dark bituminous oils and in asphalt itself. So everywhere it can be seen that the life of the Earth has descended from the surrounding atmosphere and come to work in the Earth.

9 *The Sedimentary Rocks*

Up to this point we have been considering the rocks that were precipitated directly out of living processes of the Earth and which underwent little or no mechanical change before hardening. If we now look at formations which give the impression that they originated entirely through the breaking down of other rocks and the sifting of the particles through the agency of water and air, we must be careful not to fall into the error of attributing all such formations to processes that we see today as weathering, breakdown and precipitation. True sandstones comprise only 0.7% of all rock, which shows how little of the primal rock has been worn down in order to build them.

To begin with we have to get a picture of how the sandstones are distributed in time throughout the various geological epochs. The first appearance of sandstone we can place at the beginning of the schist period. At the same time the first true limestones appear. But neither limestone nor sandstone are characteristic of this period. By the time of the Coal Measures thick limestones appear – the carboniferous limestones – and there are sandstones lying between the seams of coal.

The first great sandstone epoch begins after the Carboniferous – in the Permian system.* It consists of the sandstones and conglomerates of the Rotliegende formation which adjoin the massive thick beds of the Variegated Sandstones (Bunter Sandstone). While the rocks of the Rotliegende formation usually have a red colour throughout due to iron oxide which together with clay or silica forms a cement or mortar between the sand particles, the Variegated Sandstones are multicoloured – grey, yellowish, greenish or red. This first sandstone epoch is followed by the limestone of the Muschelkalk. In the Keuper formation which follows the Muschel-

*In this paragraph it is the German succession which is described. In Britain the earlier sandstones are equally important. (Translators' note).

kalk we have a second sandstone epoch which gradually comes to an end in the blacksand of the Jura. Following the reappearance of limestone in the white Jurassic, we find sandstone appearing for the third time in the Cretaceous and Tertiary deposits (Quader Sandstein).

This rhythm between sandstone and the residues of animal-organic processes is a true picture of the far-reaching and ever-changing life processes of the Earth. While in the limestones as well as in the slates and old granitic rocks we have a direct 'fall-out' from a life process which creates the rock, we see for the first time in the sandstones, conglomerates and marls the effect of mechanical forces which *break down* and disintegrate these old rocks. We should classify as true sedimentary rocks only those which are clearly the result of the mechanical breakdown and grading of various formations. A closer look at sandstones and conglomerates (cemented masses of pebbles) will enable us to understand better the origin of these rocks.

Sandstones consist basically of quartz granules, slightly worn but mostly sharp edged, which are cemented together. The binding material of these grains of sand is either a form of silica, clayey matter, calcium carbonate, dolomite or – more rarely – iron oxide. The binding matter of conglomerates is equally varied. The nature of the binding is not dependent on the age of the formation. The various sandstones – silicious, clayey, calcarieous – can be present from the very oldest epochs up to the Tertiary, and one often finds many kinds of binding within one formation – for instance the Bunter Sandstone (Variegated Sandstones).

In almost all sandstones fine flakes of mica are more or less visible. The only exceptions are a few pure silicious sandstones which consist of up to 98% quartz grains bound with silica, the remainder being iron and a little water (quartzites).

The mica flakes and the clayey binding indicate that the rock is the product of the debris of earlier crystalline rock such as granite or gneiss. The great masses of the Permian and Bunter Sandstone and their conglomerates usually lie immediately above the ancient crystalline rocks from which they originate.

The question now arises, what was the nature and hardness of the rocks whose debris formed the sandstones? The sharp-edged grains and scree in the breccia show that their source rocks were relatively hard. Conglomerates have a similar origin but the shape of the grains of sand and the pebbles indicate that sometimes the source material was less hard and more easily worn by the action of water. This phenomenon is a striking confirmation of what Rudolf Steiner describes as the condition of the Earth shortly before and shortly after the separation of the Moon in his lectures to the workmen September 1922. There he speaks of the viscous material of the gradually solidifying Earth, sometimes densifying to the consistency of a horse's hoof and then dissolving again. Thus there was an alternation of densities, and the solidifying of the rocks was not a linear process but a rhythmic occurrence which progressed towards the solid state.

In this alternation something of the life of the Earth as a whole comes to expression. It can be seen that even material such as grains of sand of the fragmented primal rocks was to a certain extent, subjected to the rhythmic ordering of the life of the Earth. The form which arises under these laws imitates the structure of the old crystalline rocks – namely a granular rock. In the previous epoch, the schist/slate period, these laws had an even stronger influence on the structure of the rock. The schists are often composed of the same substances as granite and gneiss. In these rocks the forces of the surrounding universe – the heavenly bodies – dominate, and, out of the soft primal magma, call forth, like an enchanted reflection, the crystals in granite and gneiss. In schist the life forces of the Earth itself determine the wood-like grain of the rock. Terrestrial forces are even more deeply involved in the formation of sandstones, conglomerates and breccias. From the *living* sedimentary forces of the schist/slate time there develop gradually the purely mechanical forces of sedimentation.

After the schist/slate period, the life of the Earth itself changed from being 'plant-like' to more 'animal-like'. This is evinced by the ever-increasing beds of limestone. The forces from the surrounding universe that prevailed at the beginning of the forming of rocks gradually withdrew and the silicious rocks are drawn into the sphere of influence of the purely terrestrial forces. This process can be traced through all its stages. We find sandstones where fractured surfaces of grains have recrystallised as if each single grain was striving to re-form a complete crystal. In the rocks of these sandstones we find beautiful nests of crystalline quartz which however usually only make the pointed tips and not the shafts as in other rocks. In the Bunter Sandstone one finds cracks lined with these pointed quartz crystals. In the sandstones of the Keuper and Cretaceous formations one has to search hard to find any. In these rocks the power to crystallise diminishes more and more, for the declining forces of the Earth take control.

We should however make a great mistake if we were to assume that the forces of the heavenly bodies in the surrounding universe had no further influence at all on the forming of rock. What we have described regarding sedimentary rocks does not apply to all other rocks.

To see why, we have to look back at the oldest geological times. The sandstones and conglomerates have a remarkable past. In old strata one finds a series of rocks some of which exhibit what resembles flattened out boulders imbedded in a uniform crystalline mass. The boulders are of the same material as the underlying rock. They are quite obviously part of the primal rock broken loose, formed into boulders and thus incorporated into the subsequent sedimentary strata. This is a very early form of conglomerate.

Or it can happen that granite or gneiss will be completely fragmented and the constituents – quartz, mica and felspar – are recemented together forming a sort of reconstituted granite. These are the so-called arcoses.

These rocks together with the Rotliegende conglomerates, the Permian Sandstones and the Bunter Sandstone occur at a period of intense volcanism and upheaval of mountain ranges. We only know of these upheavals from folds which usually lie hidden deep in the Earth but can be discovered in mines and sometimes when they are exposed on the surface. The mountains of these ranges have long since been worn down and we find their fragmented remains in the various sedimentary rocks we have described. The 'rock' of these mountains was not as hard as our present rocks. They were still 'waxy' as Rudolf Steiner pointed out. They were mostly broken down by water. The breakdown was not so much a splintering, as a crumbling, for the crystallisation process had only just begun and the various crystals were not yet firmly interlocked. The fragments were suspended in a kind of mire which was quite thick and contained a lot of material in solution. In early times, what solidified out of this mire became dense rock like quartzite for instance, where one cannot see that it consists of minute particles welded together. There exist all gradations from quartzite to the true sandstones where grains appear as if they had been hard and solid from the beginning.

Such mountain ranges like those before the Bunter Sandstone times arose again later. The younger ranges are well preserved and can be recognised by their sharp and rugged outlines. The rumps of the old mountains are either buried or else rounded and worn like the Erzgebirge, the Bohemian and Bavarian Massifs and others. Younger ranges are the Alps, the Caucasus, the Himalayas and, in South America, the Cordilleras and the Andes.

These younger mountains show that the rock masses when they rose up from the depths were still in a condition similar to the original condition of crystalline primal rocks such as granite and gneiss. The younger rocks were obviously still in a very plastic condition since they show beautiful folds in their schists like the older rocks. (Plate 1). The strange thing is that some of the younger rocks have been formed in such a way that they have the appearance of very old rocks. In Europe for instance there is a slate from the Black Jura transformed into real mica-schist, while the usual Black Jura slate, as found at the foot of the Swabian Alb, has a foliated crumbling structure. For a long time this transformed slate was mistaken for old crystalline schist until it was discovered to contain the same fossils as the Black Jura.

The transforming of a young formation into the structure of an old rock was explained as due to the enormous pressures exerted during the folding of mountain ranges. It is more likely that the condition of the rock was far more plastic and alive during the forming of these young mountains. The crystallising forces proceeding from the surrounding cosmos could then work upon them, and this is born out by the fact that the largest known rock crystals have been, and are still, found in these younger mountains. If these gigantic crystals had been formed before the folding of the mountain ranges they would have been crushed and ground to unrecognisable fragments.

So we see that even in fairly recent times, in certain parts of the Earth, rock was still not so hardened that it could not respond to these cosmic forces. In the regions of the younger mountain ranges the Earth experienced a certain enlivening which affected all rocks in the neighbourhood.

We cannot here attempt to describe all the true sedimentary rocks. Nevertheless there are still some remarkable phenomena that allow us a glimpse of an earlier condition of what we today call 'rock'. Here are two examples. In the province of Minas Geraes in Brazil and near Delhi in India there occurs a rock called Itacolumite or Flexible Sandstone. Outwardly it appears like a slaty sandstone with the pecularity that quite thick slabs can be bent to a certain extent without breaking: If one studies it microscopically one discovers that it is composed of very irregular large and small grains of quartz that are precisely interlocked. Between the interlocking grains lies a thin film of a talc-like material which does not cement the grains but acts like the fluid or film of oil in an accurately fitting joint, and facilitates the movement. In this Flexible Sandstone one finds diamonds, native gold, specular iron (iron-glance) and magnetic iron ore.

Here nature has left a record of a previous state of matter. There is no doubt that quartz grains were so plastic when the rock was formed that they mutually shaped one another as we see them under the miscroscope. However the density was such that they did not mix with the talc-like substance but this remained in the interstices between the soft quartz grains.

The second example shows that even younger rock formations such as the Flysch of Alpine Foreland regions contain varieties of rock that were not yet hard. The Flysch is a conglomerate that originated from a great mountain range that was being largely eroded before the folding that produced the Alps. This rock looks like a coarse concrete of river-worn pebbles tightly cemented together by a calcareous sandy binding material. The Flysch consists of an enormous variety of rocks from granite and gneiss to schists, limestones, sandstones, etc. In the calcareous and clayey pebbles one often finds indentations into which neighbouring pebbles fit to perfection (Plate 13). Some of the pebbles had not become as hard as they are today while others were hard enough to cause impressions in the softer material.

In the argillacious rocks and the marls, which are as variable as the sandstones and conglomerates, we can study all the transitions from true slates to actual mud deposits. In the argillacious rocks aluminium silicate, which is the basis of slate, predominates; in the marls this is mixed with lime or dolomite. Here also substances are slipping out of the hold of the living form-giving forces and are falling under the dominion of gravity which works purely physically in the depositing of sediment. The structure of these 'rocks' if one can so call them, tells us clearly what forces have been at work on them. The well-formed crystals in them show that materials were still available in which the forces of the heavens could work. Fine, flaky foliation speaks to us of the vegetative life of the ancient Earth, while dust and debris, covering the most beautiful crystals of gypsum and iron pyrites, point to the death forces of gravity and weight that take over the world of rocks.

10 *Volcanism and the Birth of Terrestrial Fire*

Volcanism – actively erupting volcanos as we know them today and from historical records – did not exist when rock began to form. As all other phenomena of living and 'dead' nature have passed through stages of development, so also with volcanism.

We have already observed that the concept of the Earth as molten primal magma is no longer tenable, especially when considering the structure of the oldest rocks such as granite, gneiss, gabbro, pegmatite, etc. We have also explained how the precursory stage of rock was gelatinous, malleable or flowing, while heat or warmth was active in it as in a living organism. From this primal condition the rock that today is called igneous underwent a gradual hardening process in rhythmic repetitions. There were times when the soft rock mass densified and other times when it again became quite fluid. This hardening process was caused by a loss of water, air and heat. By re-absorbing water, air and heat the rock was reliquified.

Such changes, which took place over the whole globe, are to be seen as life processes of the Earth. Through the nature of this more or less living condition of the rock mass, it took on forms as it flowed, which can strangely resemble the flowing lava forms of active volcanoes today. On the other hand it is evident that these moving masses of granite, greenstones and porphyry never developed into volcanoes, but only partially penetrated the soft masses of the overlying 'rock' and gradually hardened. In this way arose the 'plutonic' laccoliths, sills and veins so characteristic of these rocks.

Another stage of the developing volcanism was that the soft mass broke through to the surface and spread over wide areas in tremendous sheets. Such sheets of basalt are frequent in Greenland. They extend from there to Iceland and to the Western Isles of Britain. Another enormous sheet of basalt occurs in India in the highlands of Deccan. It reaches a height of 1300 metres and covers 10,000 sq. kms. The structure of this rock shows how thick and viscous the original mass had been. In Greenland one finds layers of basalt piled up stepwise into steep domes hundreds of metres high. Another phenomenon which also shows the viscosity of the escaping rock flow is that sometimes the mass does not manage to overflow from the crater or fissure but hardens there so that enormous, steep domes of basalt are formed. In these old Plateau Basalts one never finds the quantities of tuff, slag or volcanic bombs that modern volcanoes produce. Slow solidification in huge columns has given us the Giant's Causeway in Northern Ireland and the famous Fingal's Cave on the Island of Staffa off the west coast of Scotland (Plate 13).

If one asks what 'temperature' these flows of rock had when they erupted, one can see from the effect they have had on rocks they penetrated that they never had the heat of a present day volcano. The neighbouring rocks of these hot streams are modified it is true – they are penetrated, silicified and the limestone has become crystalline, but pieces of country rock caught up and carried along have not been melted. A low temperature is also indicated by the percentage of water contained in all such rocks today. Of course, the neighbouring (country) rocks themselves were not yet hardened and were far more subject to the action of heat and water. It was found that when such a flow penetrated a coal seam, changes took place that can be reproduced experimentally at about 500°C. But the actual melting point today of such a rock flow would be 1100°C! Obviously, present-day physical and chemical properties do not make it possible to determine indisputably the real temperature.

The solution of this problem is that with this ancient 'volcanism' we are dealing with a state of matter which today no longer exists, or perhaps only at a certain depth in the Earth. The essential thing is that there was a much more intimate union of mineral matter with water, gases and heat in the flowing rock. The union of the four elements was due to the much greater 'life' of the whole Earth. The flowing of the soft rock was a life process of the Earth organism.

Organised interplay of the four elements is a characteristic of every warm-blooded creature. In the higher animals and in man the mineral parts (bones and salts) water and gases are maintained together with warmth in a living union. None of these four elements alone can constitute life.

There are present-day volcanic phenomena which show that the magma which bursts out of volcanoes is in a condition that we can only approximately reproduce artificially in the laboratory. Measurements taken by the American geophysicist A.L. Daly in the lava lake of Kilauea on Hawii gave the surface temperature of the boiling lava as about 1200°C, but at about eight to ten metres below the surface of the molten lava it was about 100°C less. The highest temperatures (1300 – 1350°C) were in the four metres high gas flames which hovered above the surface of the lake. The burning gases contained about 60% water, 10% carbon monoxide, 3% oxygen the remainder being nitrogen and carbon dioxide. The average temperature of the molten lava in the interior of the lake was about 1050°C. Further measurements in an African volcano yielded similar results.

The individual ingredients of such a lava consist of olivine (a magnesium silicate), calcium felspar, hornblende and other minerals. Their individual melting points lie between 1900° and 1200°C, but through the high water and gas content the melting point of the lava is appreciably lowered. Such an intimate interpenetration of the four elements exists there that one must realize the very special condition of this lava. The latent heat becomes *free* when the lava comes in contact with the atmosphere. The original condition is broken up, heat is released, steam and gases escape. Another significant point is that the molten lava contains many ferrous compounds which on contact with the air are transformed into ferric compounds thus creating more heat. This is responsible for the day or even week-long afterglow of lava streams observable in many volcanoes after an eruption.

If one attempts to reproduce artificially in the laboratory such molten rock with its content of water and gases, it can only be done if the whole is heated in a closed vessel and an enormous pressure built up. Under such pressure the rock actually melts at about 1050°C. If, however, in nature such a pressure existed in a lava lake, the molten lava would be hurled up some hundreds of metres into the air. But the lake remains comparatively calm with jets rising to only a few metres from which the wind draws out the lava into glassy threads. The native Hawiians call these 'Pele's Hair'.

Such 'simmering' lava lakes have been observed at times in Vesuvius and other volcanoes. The heat produced is unimaginable.

It has been estimated that the lava lake Halemaumau on Hawii produces 300 million calories per second. – Where does this heat come from?

The second phenomenon we would like to mention is the eruption of a heavy 'emulsified' mixture of red-hot rocks and gases with no molten lava. This mixture can be so heavy that it can rush down the mountain side like a lava flood, destroying everything. It was such a fiery eruption which destroyed the town of St. Pierre, Martinique, with its 26,000 inhabitants on March 8th 1902. An impeding lava plug in the crater of the volcano Mt. Pelé caused a lateral explosion and a black cloud of ash and stones, pierced by lightning, burst out and poured down the mountain side. With a speed of 150 metres per second (360 m.p.h.) this hurricane cloud leaped on the town 5½ miles distant which was completely destroyed in a few seconds. The effect of the heat indicated a temperature of about 800°C. The neck of a wine bottle was bent like a heated candle, iron tanks were shot through by stones. Wooden planks were driven through tree trunks. The only survivor was a prisoner in an underground cell.

Such 'clouds' of glowing materials show a condition which contradicts all physical laws. Indeed they are a sort of 'emulsion' of solids, gases and heat. This mixture behaves sometimes like a liquid, sometimes like gas, at other times like a solid body. Nevertheless it is a unity and it even has its own thunderstorm centre. The lightning phenomena familiar in other eruptions show that in the vicinity of a volcano at the time of an eruption there exist peculiar atmospheric conditions. Herewith we come to an important observation of Rudolf Steiner's regarding the causes of volcanic eruptions. In two lectures to the workmen (2.6.23 and 18.9.24) he describes volcanic activity as dependent on forces which act from outer space, due to the Sun and certain angular relationships of the celestial bodies. He illustrates this by the well known phenomenon of the solfataras north of Vesuvius which begin to steam when a piece of paper is lit near them. The warm air from the burning paper produces a slight reduction of atmospheric pressure over the surface of the ground so that the gas begins to emerge more strongly. A local reduction of pressure, brought about by astronomical relationships, is the real *cause* of an eruption. Because of this it also happens that a local thunderstorm and downpour of rain occurs over the erupting volcano even if the sky is cloudless. In the second of these lectures Rudolf Steiner comes out against the theory that the interior of the Earth is molten, and points out its incompatibility with the specific weight of the Earth. The Earth is too heavy to be molten in its core.

According to this, one can picture that concentrations of heat arise which, after a long incubation, are released by a particular relationship of celestial bodies. These fire processes arise locally at particular spots. Such an explanation of course contradicts the concepts and theories of modern science which make of every fiery crater a blast furnace heated from below.

One is really dealing with an organic conception of this natural phenomenon. It sees the Earth inter-related with the surrounding world. In this sense one can consider volcanism in its present form as a life process of the Earth. The *old* volcanism was quite different. It was also an expression of the life of the Earth, but it did not have the fiery character of today.

The awakening of the terrestrial forces of fire took place very late. What we have described at the beginning of the chapter applies particularly to the younger igneous rocks such as trachite and basalt. They are transition stages (trachite) which in the beginning were still very much like granitic rocks. The trachites are light coloured and rough textured rocks composed of felspar, hornblende and some magnesium mica. The Siebengebirge on the Rhine is almost entirely trachite. The compounds of this rock often hold sufficient amounts of water to show that it was never in a molten state. Even though trachite is not so widely distributed as basalt, yet there are numerous forms and transitions which connect it on the one hand with the younger porphyries and on the other hand with the basalts which follow it. There is also the well-known phonolite or clinkstone which is so dense that it has a metallic ring when struck.

In the incredible variety of these young igneous rocks (which increases with the basalts and present-day lavas to astronomical figures) one has a phenomenon comparable to that of the lamprophyres, etc. which we have spoken of in connection with the metamorphoses of granite. To describe all these varieties would exceed the scope of this book. We can only select those which have a particular bearing on our subject.

The pale siliceous and iron-deficient trachite represents a transformed repetition of the granitic or older granular rocks.

What is typical of basalt is the much lower silica content and the generally high iron content. Because of this, nearly all basalts are dark or even black. Their mineral content, which in the main consists of augite, calcium felspar (labradorite) and magnetic iron ore, brings these rocks into a close relationship with the greenstones (gabbro, diabase, serpentine, etc.). This is shown in a very striking manner. We remember that the greenstones are associated with the plant-like stage of the Earth life. We find in them many mineral forms reminiscent of wood, for instance in the fibrous asbestos. In basalt this fibrous structure is magnified to gigantic dimensions and produces basalt columns (Plate 14). The five to eight sided pillars which can be up to 100 metres high have nothing to do with crystal formation though they are forms born of forces acting on the Earth from outside. Even as plants are in a certain sense drawn by their stalks out of the Earth by the power of the Sun, so likewise are the thousands of stem-like pillars of the basalt formations drawn out by the power of the Sun. Imaginatively one can think of the basalt pillars as the saurians of the rock world. They appear mainly at the time of the saurians (Cretaceous and beginning of the Tertiary). Rudolf Steiner has described the formative *process* of basalt as a freeing of the Earth from the 'excessive' Moon forces. The powers to effect this freeing proceeded from the Sun and entered into the rocks which were formed at the time when the Sun had its strongest influence on the Earth. These rocks appeared when the Moon was still part of the Earth and thus have become the densest and toughest of rocks and contain the hardest and most fire-resistant minerals and metals. In the partial transformation of the old rocks into basalt – and especially columnar basalt – by the Sun, lies the secret of the freeing of the Earth from the 'excessive Moon forces'.

Following the basalt, there developed gradually, in the course of the Tertiary up to the Ice Age, the lava of present-day active volcanoes, in which the original close 'organic' union of the elements has *fallen apart*. The terrestrial fire was thus born out of the world of rocks just as the great water masses were born out of the misty atmosphere of Atlantis, and descending, engulfed that continent and caused the tremendous cooling of the Ice Age.

In the lavas one can recognise certain basic types. They are primarily either trachitic or basaltic and thus show that the lava *process* developed out of these previous processes. But in actual fact there are as many different lavas as there are active volcanoes. This is due to the fact that now the terrestrial fire really takes hold of *all rocks* and transforms them. The result of these transformations are slaggy, porous materials such as pumice, innumerable forms of lava from ropey to blocky, ashes, scoria and volcanic bombs. Nearly all these materials are more or less porous and cellular, sometimes grotesquely so. The innumerable cavities in lava are due to the escape of gas from the viscous, molten mass immediately after its ejection or out-pouring from the crater. The spongy structure of the solidified lava shows that the viscous mass was full of gas, and, on the hardening of the flowing lava, rock appeared whose cavities are distinctly reminiscent of those in melaphyre and some porphyries with their beautiful agates and embedded minerals. These latter structures, reminiscent of organic forms created by the living Earth with the processes of heat, air and water, can no longer be formed in lava; the terrestrial fire has become the enemy of life.

11 *Natural Radioactivity and the Process of Germination*

With the sedimentary rocks we have been studying a stage of development that we can only describe as one of breakdown and disintegration. This process which we can still observe daily in the mountains, and hourly in the lowlands in our streams and rivers, embraces the second half of the total history of our planet. We have tried to sketch in broad outlines the first half of this history as the *building up* of the past life-stages of the Earth. This process ceased some tens of thousands of years ago and we find ourselves now in the age of the Earth which in man corresponds to the years just past the middle of his life. The middle of man's life is around his 35th year. From this time on declining forces begin increasingly to take effect. This however is necessary to enable him to achieve his full human maturity. If he has not achieved physically and spiritually by then what he should have, in order to develop it further in his later life, he can no longer make it up later. As the physical growth forces of youth begin to wane, it is only what man has been able to transform into spiritual qualities that allows him to mature rightly.

As man from middle life onwards has consciously 'taken himself in hand' in order not to get 'stuck in a rut', so he must learn consciously to counter the dying forces of the Earth in order to preserve it for some time still in a habitable condition. We need not fear an ice or heat death for the ageing Earth: the future of the Earth runs absolutely parallel to human development. When finally the Earth reaches whatever end is in store for it, then man as a spiritual being will have freed himself from his present earthly needs and will be able to pass on to a different state of existence. At present however man thinks very little about countering the disintegrating forces of the Earth. On the contrary he makes increasing use of these forces. What, for thousands of years was done only by the weathering of the rocks, he carries on by mining, deforestation, altering of water courses and unnatural methods of plant and animal husbandry. 'The Rape of the Earth' is what the sociologist and economist Alfred Weber once called our present world. Man took the last, very dangerous step towards the disintegration of the Earth when he began to make use of the powers of the disintegration of matter in radioactivity.

We do not mean to criticise the achievements of science and technology because they act destructively on the life of the Earth. Such advance is necessary because from the negative results of this 'progress' we ourselves must learn to act positively. In order to do this we must ask what really is natural radioactivity.

There can of course be the most diverse answers. We shall attempt to find one that can be read from the natural phenomena themselves, without reference to any existing theories on the nature of radioactivity. Naturally there are other possibilities but we shall leave these aside for the present.

Among the many chemical elements there are a few which, for some 50 years, have been known to give off various radiations. These invisible radiations produce either electrical phenomena (Beta and Gamma rays) or are perceptible through light effects caused in certain other substances (Alpha particles). The principle elements which give off these radiations are uranium and thorium. Also for example, the wide spread element potassium (so necessary for life) and the rare elements rubidium and samarium have radioactive properties, though very much less than uranium and thorium.

The radiations of these elements are accompanied by the breakdown of the original element which is transformed into a series of radioactive substances of varying length of life and (in the case of uranium and thorium) ends up as lead. This end product of the breakdown is *no longer radioactive*. A further end product of this uranium and thorium breakdown is the inert gas helium which likewise is no longer active. The radioactivity of these elements i.e. their continuing breakdown, cannot be influenced by any process of chemistry or physics. One can heat a piece of uranium to melting point, or cool it to minus 200°C, or subject it to a pressure of 1000 atmospheres – its breakdown is not hastened nor retarded, neither can it be halted by any means.

These strongly radioactive elements uranium and thorium are

native to the oldest primal rocks of the Earth – the granites and pegmatites. In these rocks they are extremely finely distributed and only in relatively few places are they sufficiently concentrated to be quarried or mined. Through the mechanical breakdown of rocks, before they had really hardened, uranium and thorium found their way into younger rocks such as the famous Carnotite sandstones of Colorado. The main sources of uranium and thorium for the production of atomic energy are however to be found almost exclusively in granites and pegmatites.

Today over 100 minerals are known containing uranium and about 50 containing thorium. In the radioactive minerals which one might call original (i.e. from which most of the others have originated – minerals like pitchblende, thorianite, thorite, bröggerite, cleveite and some others) there occurs something that is only found in radioactive minerals. The crystals, whether embedded in rock or free-standing, while outwardly appearing fully formed, within have the appearance of solidified jelly-like or pitchy substance (hence the name pitchblende). If, with the help of X-rays, one makes a diagram of the lattice of these 'crystals', it transpires that in their interior there is *no crystalline structure* present. In mineralogy these crystals are termed isotropic. If one heats such isotropic crystals they glow and the lattice, that according to the outer form should be there, now appears under X-ray examination.

We have the remarkable fact that these minerals, which once upon a time were crystalline, through their radioactive decomposition became jelly-like or colloidal. In fact these radioactive minerals, like all minerals, originally crystallised out of a gelatinous condition. This can be recognised from the outer form of the mineral. Radioactivity reversed the process and they reverted inside to an amorphous gel. The fact that such isotropic minerals when heated re-establish a crystalline structure is further proof of the untenability of the primal molten rock theory.

The occurence of radioactive minerals in the primal rocks and this isotropism which we have just described, are, in our opinion, important basic phenomena which lead towards the understanding of *one* aspect of radioactivity.

In order to do so however we must turn aside to an apparently unrelated subject – the *process of germination* in the plant, so that by looking at present day life-processes we recapture something of the living Earth of the past.

Seed formation is an extraordinarily involved process, part of which resembles a sort of mineralisation. When the seed in the ovary has reached the stage when the whole of it is still green and the ripening process is just about to begin, it is called 'milky'. The first garden peas we enjoy in the Spring are in this condition, as is also the 'Grünkern' of Southern Germany, made from unripe spelt-wheat. The unripe milky seeds are not viable if they are harvested and dried, for they lack a vital process which only takes place during ripening. In the milky seed the embrio seedling is fully formed, but the nutritive substances which accompany it in the form of starch, protein and fat are still in the colloidal or jelly-like state. On examination of such a seed it is easy to determine that it contains as yet very little mineral matter. During the subsequent ripening there is not only the obvious drying of the seed, but starch and protein begin to form and gradually harden out of the jelly-like colloidal state. Starch forms typical starch granules, with their fine inner layers and round or many-sided shapes. Protein congeals to real crystals which are known as crystalloids. (For example the aleurone granules of our cereals whose nutritive value is so important.)

Together with this drying and ripening, something else occurs. Starch and protein bodies become impregnated with certain minerals such as calcium, magnesia, potash, phosphoric acid, silica, etc. One can thus characterise ripening as a sort of mineralising of the seed. Not only its substance but its whole inner structure becomes rock-like or earthy. This is necessary because the seedling which remains unaffected by the drying process will later, when it germinates, need this 'earthiness' of the seed. Every seed, in ripening, builds for itself its own little Earth which supports the early life of the seedling until it finds its union with the greater Earth.

If one sows a ripe seed in Spring it begins at first to swell under the influence of warmth and moisture and, after a few days, the first thing to appear is the radicle. Before the radicle pushes through however, something important has taken place in the nutritive substance of the seed. Starch and protein have changed their structured mineralised condition and reverted to a jelly-like state.

The new jelly-like state of the germinating seed bears a great resemblance to the milky state, but is not this time followed by ripening. The nutritive substance of the seed disintegrates during germination, into water, carbon dioxide, ammonia and salts. We will not discuss here how the intermediate stages of this breakdown affect the seedling.

This gradual process of disintegration of nutritive substance and the associated root growth is accompanied by something that is not outwardly visible. *The germinating seed sends out radiations.* The existence of these germination and growth radiations was discovered by biological and physical tests. The first to describe this accurately was the Russian scientist Gurwitsch. These radiations are closely related to ultra-violet rays in that they can pass through quartz crystals but not through ordinary glass. They have a favourable effect on the growth of other living organisms and plant organs. It is probable that the traditional custom of attaching a grain of corn to a cutting to be struck is not only connected with what are known as growth substances (auxins) but also with these germination radiations.

Thus we have, in the processes taking place between the 'milky' state and germination, a true reflection of these other processes that we encounter within the mineral world in natural radioactive compounds.

If we remember what was said in Chapter 3 regarding the life processes that led to the formation of granite and kindred rocks, and then consider this ripening and germination of seeds, a very significant picture emerges. Following Rudolf Steiner's suggestion we tried to show that these primal rocks originated out of the tremendous, though at that time undifferentiated, flowering stage of the whole Earth-being. Looked at in this way we might say that the nature of the old crystalline rocks has something in common with a seed. If one considers that these rocks constitute something like 90% of the mass of the Earth, the idea can arise that the Earth as a totality is a giant seed. This 'seed' Earth ripened during the time which elapsed between the first gel condition of the rocks and their present solid state. This hardening of the rocks is however a very gradual process and at different times and in different places alternates with softer conditions. There is evidence that at the times of the great mountain range upheavals the rock mass in certain areas was still in, or had reverted to, a pliable condition; for otherwise the crystals and minerals in the hollow cavities would have been completely crushed. Further evidence of late hardening-up of the crystalline rocks is to be found in the cultic buildings of earlier civilisations where for example in Egypt and Peru, *without iron tools,* stones were accurately dressed from basalt, diorite, granite, etc. These buildings date from between 3000 BC and 1000 AD.

Hardening, together with mineral formation, continued right into the Christian Era but the processes of disintegration, weathering and mechanical break-down of rocks began much earlier. It began from the time when no further *new* rocks were formed as a result of *living* processes.

The minerals which today are radioactive were present in the primal rocks during this whole process of hardening. They are evidently a sort of ferment which, extremely finely distributed, works towards the dissolution of the 'Earth as a seed'. One might say that the 'Earth seed' is beginning to 'germinate' so that out of this dissolution a new future existence may arise. We will not discuss here the nature of this existence. In Rudolf Steiner's works there is much to be found about it.

We will pass on to something else. From Rudolf Steiner's investigations it appears that before our present epoch there were two other great cultural epochs which have disappeared without trace. In the older of these, the Lemurian epoch, the conditions of existence were completely different. Man still had the capacity to influence the form of the animal. He had power over the reproductive forces of animals. The Lemurian epoch disappeared through the misuse of these powers. In the succeeding Atlantean epoch, on a continent between Europe and America, man had the power to work creatively on the plant world. He had control over the germinating and growth forces of plants which he could use for technical purposes. The Atlantean continent and its civilisation perished through the misuse of these forces between the 11th and 8th millenia B.C.

We have to thank these two prehistoric epochs for our domestic animals and food plants. Our own culture is based entirely on the mastery of the mineral kingdom.

Now, however, after the discovery of atomic energy, which is based on the natural radioactivity of certain minerals, our epoch begins to make use of the equivalent in the minerals of the 'powers of germination.' When these forces become misused then our civilisation too will perish. By the time this happens man must have reached new levels of existence. Disintegration, destruction and death are the foundations of new life.

12 Meteoritic Phenomena

Life in our present day Earth is no longer active in the world of rocks but only in plants and animals and in a certain way in air and water, but there is one way in which it still enters into the world of rocks and metals – the formation of meteors in the atmosphere.

Before we go into the nature of this process we must be quite clear about what is meant. Meteoric phenomena are usually understood as comprising all 'falling stars' i.e. shooting stars, meteors and fire balls but excluding comets whose behaviour is altogether different.

Astronomical research on meteors in recent decades, especially the work of Hoffmeister (Sonneberg, Thuringia) has shown that among all the directions in the whole heavens from which shooting stars, fire balls and meteors proceed (radiants), there is one stellar region from which come nearly all meteors that reach the Earth as meteoric stone or meteoric iron. This region lies in the constellation of Scorpio.

When, around August 12th of each year, we see the shooting star showers of the Perseids, we must understand clearly that these are *light* phenomena from which no meteoric stone or iron falls to the Earth. The same holds good for all shooting stars during the course of the year.

The fall of meteorites (stone or iron) is always accompanied by loud noises (explosive thunder, rushing and hissing) and intense light phenomena and, in the case of fireballs, the forming of a tail. The thunder clap which accompanies the extinction of light of the meteorite is not due to an explosion of the meteorite, but rather to the *implosion* of a luminous space. What later forms the metallic or stone meteorite enters the atmosphere as light (physicists would say in a highly ionised state or 'plasma') and in the moment of implosion suddenly condenses as solid matter. An observation of the

Treysa meteorite near Cassel on April 3rd 1916, recorded a fireball of 1000 metres diameter at a height of 50 km. which rapidly diminished to 400 metres and at a height of 16 km. imploded, and dropped as an iron meteorite of 63 kg. with a diameter of 36 cm. The shower of meteoric stones at Pultusk in Poland in 1868, which scattered some hundred thousand stones, originated from a light phenomenon 300 metres in diameter at a height of 50 km.

Thus it appears that a huge body of light suddenly condenses into the solid state, rather as in a case of sublimation i.e. the transformation of a gaseous substance into solid crystals. The impression of this sudden solidification is to be seen in the structure of meteoric iron and meteoric stones. In the case of meteoric iron there is an elaborate complex of large crystals in the interior of the meteorites (Plate 11) built up of varying proportions of iron and nickel. This Widmannstetter structure cannot be produced artificially. In fact if one heats a piece of meteoric iron showing Widmannstetter structure this will be completely destroyed at about 900°C and what remains will be indistinguishable from ordinary terrestrial iron. This tends to prove that meteoric iron was *not* molten before it hardened. It originated from quite a different condition of substance than is known to us on Earth.

In the case of meteoritic stone the sudden condensation is suggested by the unique texture of the chondrites. These chondrites contain small globular bodies called chondrules ranging from microscopic to pea size, which often have an inner fibrous radiating structure. They are embedded in a ground mass of the same composition. Other chondrites are not crystalline in structure but resemble tuff i.e. they are porous. *The phenomenon of chondrites is entirely confined to meteoric stones and does not occur in any terrestrial rocks.*

Such chondrites are really a sort of emulsion of finely distributed iron and other minerals which can only have been formed in a non-gravitational field. The heavier and lighter constituents did not separate according to the law of gravity, but remained mixed like an emulsion of water and oil. By contrast, although basically every granite, syenite or diorite is made up of heavy and light minerals, in these rocks one *never* finds chondrules which are so characteristic of the meteoric rocks. Further, in these terrestrial rocks one finds only intergrown crystals of various minerals. It is precisely the forming of chondrules which gives the clue to the rapid condensation into drop-like forms.

With shooting stars this whole process only reaches the stage of luminosity, but the showers of meteorites from the constellation of Scorpio reach the density of matter. This does not mean that no material substances ever reach the Earth from shooting stars. But their individual mass calculated from their light and velocity amounts to no more than a fraction of a gram. It may be rather rash to draw conclusions regarding 'mass' from pure phenomena of light and velocity in space, but it is likely that extremely finely distributed 'shooting-star substance' exists in the Earth atmosphere.

Shooting stars and meteoritic showers differ in other respects. Shooting stars have a lower velocity – average 41.6 km. per second – and an almost parabolic path, while the large meteorites reach the Earth at over 62 km. per second following a hyperbolic path. Furthermore most of the shooting star showers have an indisputable connection with disintegrated comets, while it has never been observed that the origin of a meteoritic shower has any connection with comets.

The maximum annual fall of meteorites occurs in June when the Sun is in Taurus. This is a minimum period for shooting stars. The number of meteorites begins to increase in April and then in July decreases to below the April figure. It is important to realise that in June the Sun stands in Taurus and it is from the opposite direction, from Scorpio, that the meteors originate, which fall as meteoric iron and meteoric stone.

This does not mean that such showers only occur from April to June, for there is a certain degree of activity the whole year round from the direction of Scorpio but it only reaches its maximum when the Sun is in Taurus. These variations in the course of the year suggest a living organic process which is no more precisely calculable than the growth of a tree.

To complete the picture it should be noted that the area of the heavens lying between the constellations of Scorpio and Taurus is notable for starless areas and dark nebulae. On the basis of observation, modern science has come to the conclusion that there must be a greater 'density of matter' in these areas obscuring the stars behind.

What do we know about the origin of meteorites? Let us begin with what can be learned from the composition of meteoric stone which is more common than meteoric iron. Most meteorites are chondrites which, as we know, are a mixture of finely mingled wisps of iron and rock.

The minerals of which this rock is composed are, strangely enough, always the same. What sort of stones are these which 'fall from heaven'? The answer is: usually diabase – a rock of the family of greenstones which are dealt with in Chapter 3. In this family of greenstones the older 'relatives' of diabase are gabbro and serpentine, the younger relative is basalt. Likewise some meteorites resemble gabbro, others basalt. They also contain a characteristic suite of supplementary minerals. Diamonds occur very occasionally, graphite more often, also ferro nickel, iron sulphide (magnetic pyrrhotite) which here occurs as troilite, magnetic iron ore and iron chromate (chromite). All these minerals also occur on the Earth in the various gabbros, diabases and basalts.

However, meteoric stone and some meteoric iron also contain minerals which are *unknown* among terrestrial rocks. There are, for instance, oldhamite, daubreelite, schreibersite and rhabdite. The principle minerals which constitute the rock of meteorites are plagioclase (calcium sodium felspar), enstatite, bronzite and hypersthene (all magnesium iron silicates), diopside (magnesium calcium silicate), augite (magnesium iron aluminium calcium silicate), olivine (magnesium iron silicate), fosterite (magnesium silicate). These are the same minerals which are also the constituents of the terrestrial gabbros, diabases and basalts. Olivine plays a particularly prominent part in meteoric stones. It was found in gemstone quality (chrysolite) in the meteorite of Krasnojarsk in Siberia.

The remarkable thing is that ferro-nickel is found in terrestrial rocks such as dunite and peridotite which consist mainly of olivine. Examples of terrestrial ferro-nickel are the awarvite of New Zealand, the josephinite of Oregon, U.S.A. and the souesite of British Columbia. Then there is the ferro-nickel found in the basalts of the Island of Disco (Western Greenland) and of the Habichtswald in Bühl near Weimar.

If one brings all these facts together it appears that in the meteoric stones there is the equivalent of the greenstones lying deep within the Earth between granite, gneiss and the crystalline schists.

Let us remember that the greenstones by their very nature and structure are closely connected with the 'tree stage' of the Earth. We also pointed out that the organic wood-like structure and the composition of these minerals – greenish magnesium iron silicates – are the 'signature' of the 'plant stage' of the Earth. We mentioned that former life stages of the Earth were recapitulated; in this case, the stage when the Earth, Sun and Moon were still one heavenly body. We then described how this stage was followed by the formation of schists and slates and that *after* the separation of the Moon-Earth from the Sun, there arose the Coal Measures. The formation of the greenstones however falls entirely in the time *before* the separation of the Moon-Earth from the Sun and *before* the formation of the Coal Measures. The basalts are later transformations of the greenstones as we have described in the chapter on volcanism.

We can say therefore that the origin of the greenstones is connected with the Sun evolution of the Earth and their diversity comes about through the threefold recapitulation of former life stages.

*

Rudolf Steiner has stated that the substance of meteorites and shooting stars was radiated out into space from the Sun. This radiation should be understood as a light process – substance in a state of light. The modern physicist would speak of a highly ionised state of matter assumed to be in space. The radiation of the light substance from the Sun is connected with sunspots. One might take it as a reaction to the formation of sunspots. The light substance which is thrown out from the Sun appears within our solar system as meteors and shooting stars.

This information, acquired by spiritual scientific research, Rudolf Steiner first spoke of in 1923. It was corroborated in the early 1940s by the work of Bengt Edlen, B. Strömgren and M. Waldmeier who, through spectrum analysis, found that the light emanating from the corona and the edges of sunspots showed both qualitatively and quantitatively the same substances (elements) as are present in meteors. (This phenomenon was attributed to meteors falling into the 'incandescent' globe of the Sun, but in reality what occurs is exactly the opposite.)

Thus it is evident that today on the Sun, 'substances' are being formed in the same way as during a past life stage of the Earth when the two bodies were one. Through the living interaction of Sun and Earth the family of greenstones came into being. What in those days was achieved gradually through long periods of time by the interaction of Earth and Sun, is today a process proceeding from the Sun and affecting the life of Earth and man.

The formation of meteoric stones is a metamorphosis of that past process which produced a particular family of rocks on Earth. How does it happen then that meteorites are mainly formed in and proceed from the direction of Scorpio? Let us follow a clue given by Rudolf Steiner when he describes how the great periods of development stand under the 'rule' of particular areas of the Zodiac. For Old Saturn it was the Lion; for Old Sun, the Scorpion (called in earlier times the Eagle); for the Old Moon, the Waterman and for the Earth, the Bull. One should picture these areas as the 'seat' of those creative spiritual forces guiding each particular stage of development. So the Sun period was under the sign of Scorpio and the present Earth period under Taurus. Thus it is from Scorpio that the forces are active which formerly 'ruled' the joint life of Sun and Earth. The 'Sun life' of the Earth proceeded from Scorpio and what was expelled from this life as 'substance' condensing right down to the mineral state, received its impulse from this region. What today radiates from the sunspots as light 'substance' is received by the constellation of Scorpio and reflected to the Earth from there. Taurus the Earth sign, lies exactly opposite Scorpio. On this line Scorpio-Taurus, right across space and our solar system, is the field of starless areas and dark nebulae of which we have spoken above. One can only understand this phenomenon if one considers that there are qualitative differences in the spatial direction within the zodiac. *The direction Scorpio-Taurus is the line along which takes place the densifying of the 'solar-light-substance' into earthly rock and metal.*

Astromomers know that the meteor showers from Scorpio are the only ones originating in interstellar space having *no* connection with comets, but cannot explain why. They know that meteoric stones always consist of greenstone materials and never of granite, gneiss, schist or limestone, but again they cannot explain why. If one brings together the findings of natural scientific and spiritual scientific research an explanation emerges and these things fall into place.

13 What is the Mineral Composition of our Rocks?

About 95% of the known rocks belong to the igneous group. It is striking that the foundation of our world consists of a particular class of rocks characterised by a crystalline, granular structure. The quantity of slates and shales, most of which as regards their basic composition resemble the igneous rocks, is negligible – a mere 3.7%. Likewise the sandstones, containing an average of up to 80% silica, constitute only about 0.7% of the total rock. What remains are the limestones – 0.2% consisting of about 80% calcium carbonate.

We will take these four main groups of rocks together and see of what minerals they are composed. Initially we will concern ourselves not with the constitution of the individual rocks but with the minerals which are the *main building materials* of the world of rocks as a whole.

The figures which follow are the outcome of many decades of research. They summarise the results of hundreds and thousands of analyses by methods able to detect fractions of milligrams of substances present in the sample.

One of the most important findings is that felspars account for well over half (59.5%) of our rocks. The next largest group comprising augite, hornblendes, olivine and mica amounts to 20.6%. After these minerals, which are all compounds of silica, come pure quartz (also found as rock crystal), sandstones and loose sand, with 12.6%. These three groups of silicic minerals together form 92.7% of the total rock mass.

The remainder, about 7% is apportioned between approximately 4% iron ores, 1.6% calcite and dolomite, 1% clay (leoss, loam, potters clay, kaolin, etc.), accompanied by widely distributed minerals such as garnet, apatite, zircon, titanium ores and manganese ores which make up the remainder. These figures appear at first very abstract and meaningless, but if we consider them in relation to past stages of the life of the Earth they begin to speak.

What does it signify that nearly 60% of rock consists of felspar? In order to relate this to our earlier studies of the mineral-plant and the plant-animal life of the Earth we must take into account the composition of this mineral. Felspar and its relatives, including the so-called felspathoids, fall into three different basic groups. These are potassium felspar – orthoclase (a potassium aluminium silicate), sodium felspar – albite (a sodium aluminium silicate) and calcium felspar – anorthite (a calcium aluminium silicate). These aluminium silicates do not however occur pure in nature, for all felspars contain *two* of these other components in larger or smaller proportions. These two components are either potassium and sodium, *or* sodium and calcium. Thus the potassium felspar (orthoclase) in practise is a potassium-sodium felspar with little sodium, and sodium felspar (albite) contains always a certain quantity of calcium. Likewise calcium felspar (anorthite) always contains a little sodium. All this goes to show that in nature there are no pure substances as man with his love of classification would like to imagine.

Between orthoclase and albite we have sanidine and the perthite felspars, and between albite and anorthite the plagioclase series with oligoclase, andesine, labradorite and bytownite. Besides these there are the felspathoids such as leucite, nepheline, sodalite, nosean and hauyne. Leucite is really a potassium felspar with little silicic acid and nepheline is a sodium felspar, also poor in silica; sodalite, nosean and hauyne are nephelines containing respectively sodium chloride, sodium sulphate and calcium sulphate.

One should not be put off by all these names which are basically unimportant, they are mentioned for the benefit of those who would like to pursue these matters in greater detail.

It can be seen that the above series of felspars, as we have described them, are an expression of the development of rocks in accordance with their underlying life processes. The path from potassium felspar to sodium felspar and on to calcium felspar is the same as the progression from oldest granites to the greenstones and porphyries, and on to trachite, basalt and lava. Here in the minerals we see the same progression as in the rocks – the increase in calcium and the decrease in silica. This represents the transition in the underlying life processes from the mineral-plant to the plant-animal and finally to the more or less pure animal life processes. If one takes into account that it is indeed the felspars which owe their origin to the old flowering stage of the Earth's life, it is clear that the plant nature, which expresses itself most purely in the potassium felspar, is gradually displaced by an animalising tendency expressed by the sodium and calcium-rich felspars.

Thus we find potassium felspar (orthoclase) mainly in granite, gneiss, syenite, some porphyries and trachite, but sodium-calcium felspar (plagioclase) in its various modifications in the greenstones, crystalline schists, porphyries and basalts. The strange felspathoids – leucite, nepheline and the sodalites – are to be found mainly in the young volcanic rocks and in recent lavas.

Such a rough classification should not be regarded as a rigid system, it is only a broad outline which, through the variability of living things, is continually modified. It is just these exceptions to the rule that are the surest sign that the rock-forming processes did not originally follow *chemical* laws but laws of life which were different from those we know today. If chemical laws *only* held sway, the world of rocks would present a much greater uniformity and simplicity.

In connection with the felspars it should be noted that their inner crystalline and molecular structure, from the point of view of mineral chemistry, has a pecularity first described by the Russian geochemist Vernadsky. The inner structure of felspar is based on a ring (the kaolin nucleus) consisting of aluminium oxide and silicic acid. This substance occurs naturally as kaolin or china clay. It is also the basis of all plastic clays, potters clay, brick clay, etc. It is significant that this plastic substance has an inner structure which is *not* found elsewhere in the so-called inorganic world. We find a ring structure notably in the substance which is the foundation of all life – proteins, and which is the most plastic substance known. This phenomenon indicates that the felspars have retained an impression of the life which cast them forth.

*

When we pass on to the second large group of minerals – the micas, augites, hornblendes and olivine – we find materials and structures with totally different forms. While in the felspars aluminium oxide plays a major role, it loses all significance in this second group of minerals. In place of aluminium oxide there appear, in the main, two substances – magnesium and iron. The transition between the first and second groups is formed by mica. In it there is still the kaolin nucleus and therefore the important substance aluminium oxide. A typical potassium mica (potassium-aluminium silicate) is the light-coloured, silvery muscovite which is practically free of iron. It is found exclusively in granites, pegmatites, gneiss, mica schists and phyllites and is often accompanied by potassium felspar (orthoclase). Muscovite is however entirely absent from the greenstones and porphyry group and the greenstone slates. In some pegmatites the mica sheets are as much as a metre across but they do not reach the tremendous size of the felspar crystals in these rocks. (We mentioned a whole quarry in the Urals working in a single gigantic crystal). Sodium mica (paragonite) and calcium mica (margarite) play only a local and subordinate role, likewise lithium mica (lepidolite) and lithium iron mica (zinnwaldite). This latter however is a transition to the coloured magnesium iron micas, biotite and phlogopite. Phlogopite is the mica which forms probably the largest crystals, for instance at Sydenham, Ontario, where crystals of 1.5 × 2.5 × 5 metres occur, i.e. giving mica sheets of 1.5 × 2.5 metres. The real magnesium mica (biotite) is a major or minor ingredient of many granitic rocks, some gneisses and mica schists as well as mica syenites and diorites, porphyries and trachites. The colours of these magnesium iron micas are always dark – brown, green or black.

Related to magnesium mica are the chlorites – fine green flaky minerals bearing many names. The thin green film observable on many rock crystals and felspars in the Alps is due to chlorite. Other minerals of somewhat mica-like structure are talc (soapstone or steatite – a porous magnesium silicate coloured green by iron) and serpentine in the form of foliated serpentine (antigorite).

There is another form of serpentine which makes a bridge to the pyroxenes (augites, etc.) and amphiboles (hornblendes, etc). This is the fibrous crysotile-serpentine whose green wood-like formation in dense serpentine is a matrix for asbestos – rock wood and mountain leather. Here we arrive at the significant point where the plant-forming forces of the whole living Earth reveal themselves most clearly in the mineral kingdom. In serpentine asbestos, as in hornblende asbestos, the crystal-forming forces of the surrounding cosmos are almost completely neutralised by the formative forces of the life of the Earth itself.

The groups of the pyroxenes and amphiboles are divided into magnesium iron silicates, calcium iron silicates, calcium magnesium silicates and sodium silicates. Most of these minerals are of a lighter or darker green colour, or even nearly black; some show a brownish tinge. Many tend to have fibrous scaly cleavage surfaces giving them a satiny iridescence like mother of pearl. One has the impression that these minerals, in the same way as chrysotile and asbestos, were moulded by organic formative forces and not much influenced by crystal-forming forces.

The magnesium and magnesium iron silicates – enstatite, bronzite, hypersthene – are typical ingredients of many gabbros, norites, melaphyres and basalts. To this can be added the calcium magnesium silicates – diopside and augite, both of which contain iron and appear in the same rocks as well as in crystalline schists and greenstones. Aegerite, a sodium iron silicate, which can form the finest hair-like, fibrous bundles, is found in granites, pegmatites and syenites.

The transition from gneiss to crystalline schists and greenstone schists is characterised by the amphiboles which include hornblende. They have a characteristically fibrous structure, often fine enough to be classified as asbestos. The anthophyllite, found near Hermansschlag in Moravia, forms striking yellowish-grey radiating fibrous bundles in mica balls, and in the Transvaal it is mined in large quantites as fibrous asbestos. Anthophyllite is a magnesium iron silicate. In crystalline schists we find actinolite, tremolite, nephrite and grunerite. Actinolite is one of the most wide-spread amphiboles with its beautiful dark green, radiating bundles of acicular cystals. It is found in the talc and chlorite schists of the Zillertal. It also forms actinolite schist and, as grass-green smaragdite together with blood-red garnet, gives the unusual eclogite rock its unique colouring. Actinolite is also the mineral that forms the semi-precious stone amiante (cat's eye) as well as the asbestos minerals mountain wool (byssolith), mountain leather, mountain cork and rock wood.

Hornblende (a calcium magnesium iron silicate) as well as occuring in crystalline schists is also found in the transition rocks between hornblende granite and true greenstones. The sodium hornblende (arfvedsonite), with an iron content of up to 30%, is sodium iron silicate which, together with the sodium felspar nepheline, forms many syenites and similar rocks. Closely related to arfvedsonite is riebeckite, whose tourmaline-like crystals point to this gemstone's close connection with hornblende. Not unlike riebeckite is crocidolite. It only occurs in crystalline schists. Its varieties include an asbestos and the two semi-precious stones tiger's eye and falcon's eye. The former is yellow, the latter a dull greenish blue.

The remaining mineral of this group is olivine, a magnesium nickel iron silicate, which independently forms rocks such as dunite and periodotite (of the greenstone group) as well as constituting a large part of gabbros, diabases, melaphyres, basalts and many crystalline schists. Olivine is one of the characteristic minerals of meteoric stones. In gem quality it is chrysolite.

It is typical of these mica, pyroxene and amphibole groups that magnesium and iron play major roles. Magnesium expresses the plant-forming forces; it is the main agent in the production of chlorophyll. Iron is present mainly in its ferrous form and gives these minerals their lighter or darker green shades.

The structure, colour and composition of these minerals taken together make it clear that plant and plant-animal forces were active in the forming of the greenstones and crystalline schists. The animal formative forces show themselves in the gradual introduction of sodium and calcium.

Iron ores constitute between 4% and 5% of the total terrestrial rock. 4% occurs as iron oxides, the remainder as pyrites and iron silicates.

By far the greatest quantities of iron oxide ores (to which belong magnetic iron ore – magnetite, titanium iron ore – ilmenite and chromic iron ore – chromite) are to be found very finely distributed in the greenstones, especially in gabbro, diabase and the basalts. Magnetite is what gives the dark, almost black colour to many of these rocks, especially basalt. Iron ores are the only metal ores which take an active part in the forming of rock and thus indicate that iron plays a role as important in the Earth-as-a-whole as in the living realm of plant, animal and man.

The increasing appearance of iron in the transition from granites to greenstones is closely linked with the early differentiation of life. Iron gives the impulse towards the individualising of forms in the kingdoms of nature.

14 *Metals in the Life of the Earth*

Whoever goes about with an observant eye cannot fail to notice that soils and rocks often have quite strong colours. Although the brown and yellowish shades of the weathered layer predominate, yet, if one penetrates deeper, all other colours are to be found. These colours are obviously not as vivid as the pure mineral colours such as cinnabar (vermilion), cobalt blue and chrome green, but on large sunlit surfaces they can lighten up more brightly. Besides the red sandstones and the striking red Keuper Marl, there is the rich colour range of the granites – blue from the Kosseine in the Fichtelgebirge, pink from Baveno in Italy, green from the central Alps and red from the mountains of Scandinavia. In the Grand Canyon of Arizona, the evening sun touches with magic the multi-coloured rocks ranging from gold to purple calling forth ever new transparent hues in the waning light. If one asks what causes these colours in rocks and stones and soils, the answer is always: iron.

Iron with its colours penetrates the whole Earth. In the old rocks it is often so extremely finely distributed in its compounds with silica, oxygen and sulphur that it appears as if a mighty cosmic breath of iron impregnated the still airy-fluid living Earth. The Earth inhaled this iron and incorporated and transformed it in manifold ways. Thus appeared the different colours. The earliest iron that the Earth took in is responsible for the greens and blues. Already in some granites, but above all in the greenstones, we have this 'green' iron. Seen from this aspect, iron is strongly akin to nickel which also gives a green colour to minerals when present in very small quantities. Then there is chromium which likewise produces shades of green in the gem-stones emerald, jade and nephrite. All three metals are found together in the greenstones but it is the 'iron process' that is responsible for their colour.

In the older greenstones, some of which resemble granite, and in some schists as well as in the basalts (descendents of the green stones) iron is found not only as green iron silicate but also in the form of magnetic iron ore. This latter contains much oxygen. It is a remarkable combination of ferrous and ferric oxide – a half 'burnt up' kind of iron. One should not imagine that this ore has arisen from red hot metallic iron. When the blacksmith draws glowing iron from the fire and forges it with his hammer, the little fragments that fly off (hammerscale) resemble chemically this half 'burnt up' iron, but magnetic iron ore (so named because in its natural state it is magnetic) has not arisen in a forge. Such metal ores in the old rocks point to living processes which formed them. The iron that the Earth inhaled was taken up by these life processes. Man today also inhales minute quantities of iron with the air. It is most prevelant in late Summer and Autumn when the shooting star showers brighten the night sky. This is why the iron content in human blood begins to increase towards winter time and in the depth of Winter contains far more iron than at Midsummer.

The iron which we inbreath and take into our blood is what makes it flow. In the marrow of certain bones are the blood-producing vessels. Small cells detach themselves from the walls of these vessels. The cells at birth are colourless and possess a nucleus. As soon as iron enters these cells their nucleus disappears and they form red blood corpuscles. This primary blood is however a sort of virgin blood. It has not been through the lungs nor the liver. It is as yet neither blue venous nor red arterial blood. The bluish venous blood is a 'plant-like' blood – it is saturated with carbon dioxide which it releases in the lungs in order to take up oxygen and become the bright red arterial blood.

In venous blood iron is present in the 'plant-related' bivalent (ferrous) form, corresponding to green iron silicate. In arterial blood the iron is red owing to the oxygen and thereby so transformed that in this breathing process a soul-endowed being can dwell. In other words, the iron has become 'animalised'. In the mineral kingdom 'red' iron is always trivalent – ferric – to put it in chemical language.

The ferrous iron oxide in magnetic iron ore corresponds to venous blood – the ferric iron oxide to arterial blood. Thus in magnetic iron ore we have a *mineral* form of iron corresponding to what, in man and animal, is the iron of the 'virgin' blood of the bone marrow. The virgin blood is neither venous nor arterial; it is the precursor of both. Magnetic iron ore occupies a similar position in the life of the old Earth; it is the origin and beginning. It is the principal iron ore and the oldest and is very finely distributed throughout many rocks.

In the same way as the newly formed iron-enriched blood enters the arterial and venous life cycle, so we find in the subsequent development of rock, the appearance of red iron stained sandstones and red iron ore (haematite). All this iron is trivalent iron oxide. It is combined with oxygen (compare arterial blood). At the same time the bivalent form appears combined with carbon dioxide in the puzzling siderite or iron carbonate (compare venous blood).

It may seem surprising that we try to connect the origin of such a mineral as iron ore with life processes of a higher order. It is however a continuation of the line we have taken from the beginning in observing the world of rocks. This particular comparison applies to iron because it occupies such a central position in all life processes. Iron is involved in the whole of organic development. The formation of its ores and all its other occurences in the mineral world have been shaped by life. Even the natural processes involving iron such as weathering and the infiltration of iron into springs and the living humus of plant-sustaining soils can be seen as life processes of the whole Earth. The Earth has inbreathed and incorporated this metal in order that otherwise invisible life should be made manifest in all its diversity of individual forms.

There are other iron ores which also play a part in rock formation. These are the sulphur compounds of iron – pyrite, marcasite and pyrrhotite (magnetic pyrites). Pyrite is mostly found in the crystalline schists and the blue roofing slates of Paleozoic time. It is also found finely distributed in granites and in fissures of the oldest rocks. This again goes to show how this heavy iron ore was suspended in the old gelatinous state and then gradually crystallized in small crystals. It also shows that iron did not rise up from the depths but that it originated from the surrounding spheres. This is also why practically no workable deposits of iron ore are to be found in the oldest rocks. Iron originally entered the atmosphere in such a dispersed state that it could not accumulate in large quantities.

There was however, in this atmosphere which we described earlier as the world of the mineral-plant, a substance which is not to be found in our atmosphere – sulphur. Had it remained in the Earth's atmosphere, subsequent higher forms of life could not have developed. Where did this sulphur come from? All protein contains a small quantity of sulphur, but the 'primaeval albumen' contained sulphur and all the minerals found in rocks today. As the old life began to die, the albumen broke up; the minerals were precipitated and sulphur went into the atmosphere. It would never have disappeared from the environment of the Earth if the metals had not come down to the Earth from 'outside' and drawn the sulphur down with them. From this process all the metallic sulphides originated. Pyrites, galena and cinnabar, etc. did not appear at once in crystalline form, but passed through the gelatinous state. Rudolf Steiner calls this the 'parent state' of the later ores. One should not imagine that the process of sulphur combining with metals was merely a chemical action in the modern sense. On the contrary, it was an organic life process of the Earth.

The whole process presupposes a condition of metals that one might describe as 'aeriform'. Where do we find evidence that in these early conditions of the Earth metals were indeed so finely 'vapourised' that one could call them gaseous? – In the colouring of gemstones and other minerals.

Rudolf Steiner describes very early Earth conditions in which metals floated round the Earth like coloured clouds and gradually condensed from this airy state. We have the history of this recorded in the colours of the precious stones. The colours are due to finely distributed metals especially *iron,* manganese and the iron-related chromium and titanium.*

Metals are rarely to be seen in their pure state except gold and copper. All other metals are not so intensely coloured as these two. However it is possible to reduce pure metals to fluid or vaporous conditions, when, as colloids, they show intense colours. In this way the metals have in a sense returned to an earlier condition.

Thus everything metallic is born of the air. If it is asked from where the metals in this aeriform state descended to the earth, we can refer to the experiments of L. Kolisko. For many decades she

*See Walther Cloos, *Kleine Edelsteinkunde*

studied the behaviour of metals by allowing solutions of metallic salts to rise up strips of filter paper during various planetary constellations (conjunctions, eclipses, oppositions, etc.). From the resulting chromatograms it was shown that certain metals belong to certain planets. It appeared for instance that when there was an occultation of Mars by the Moon (a conjunction), the solutions of iron sulphate and silver nitrate behaved differently and showed different pictures from those obtained when the two bodies were separated in the heavens. Thus it was possible to determine that lead was connected with Saturn, tin with Jupiter, iron with Mars, gold with the Sun, copper with Venus, quicksilver with Mercury and silver with the Moon. Some of these experiments have been reproduced by other researchers, These experiments show that when the metals are in solution they again come under the influence of the appropriate planetary sphere.

Solid metal or metal ore does not reveal its origin. It is so to speak frozen and rests in the ground like a seed. But the moment it is dissolved as a clear solution or a colloid, it becomes *sensitive* to the *planetary sphere* to which it belongs. If conditions in this sphere change – visible as astronomical planetary aspects – then it appears that variations also occur in certain chemical characteristics of the appropriate metal solutions on the Earth. Thus, for the first time, the correspondences between metals and planetary spheres have been demonstrated scientifically. The connection between metals and planets has been human knowledge through the ages. For the old Indians, Persians, Egyptians, Babylonians and Greeks it was a fundamental fact. Rudolf Steiner confirmed it from his own research and encouraged experiments of the kind undertaken by L. Kolisko where these things could be made manifest.

How could it have happened that metals could reach the Earth in an aeriform state since the various heavenly bodies of our planetary system are so distant from the Earth? Let us remember how in earlier times and under past conditions, the Earth was far larger and contained within it all the other bodies of our solar system. At the beginning of the 'Earth Evolution' this combined body occupied a space extending to the orbit of the present Saturn. At this stage it was a pure warmth organism, which however had begun to differentiate. When this stage was completed the Earth body gradually contracted to the present Jupiter orbit and left the planet Saturn behind as a separate body, and the Earth became a planet of heat, air and light. As this second stage drew to its close another shrinkage took place which left Jupiter behind on its orbit. At a third stage Mars was left behind. Later on in the course of the Earth's development the Sun was separated and the inner planets Mercury and Venus separated off from the Sun-body. Meanwhile the Earth had shrunk to approximately the orbit of the Moon. Only after the Moon also withdrew from the Earth could that layer of rocks and minerals arise which is the subject of this book.

We have here briefly sketched the cosmology of our planetary system. Rudolf Steiner has elaborated it in detail in the two books previously mentioned – *Occult Science* and *Cosmic Memory*.

The extraordinary complexity and diversity of the creation of the world can only be touched on here. We have introduced it in order to show that the planetary system is an organic development and not a spiral nebula spun by an imaginary deus ex machina in order to throw off the planets as glowing clouds of gas as maintained in the Kant-Laplace theory. We present this picture of an organic cosmology in order to illustrate how metals in their primal form were borne to the Earth from outside in an aeriform state; how the withdrawing planets left behind, as it were, seeds in the Earth, and these seeds were and are the individual metals. Seen in this light the metals are gifts of the various planets and their existence in the Earth is like a memory of past stages of life.

Naturally many objections can be raised. One might say that if the metals have been built into the Earth in such a beautifully ordered sequence, then they should also be found in an orderly fashion in the Earth's strata. Another will say that there are far more metals than the seven here mentioned as connected with evolution.

To the first objection we would reply that this could not have happened since the Earth was and is a *living* organism and the metallic nature was incorporated as a *living* and not a mechanical process. The Earth had to digest the metals so that these gifts of the planets could be used by the life processes. Through the absorption of metals there arose an organic growth as in plants or animals, not a mechanical arrangement.

To the second point we would reply that the seven fundamental metals represent something like the seven notes of a scale. As there exists a great variety of intermediate tones within the scale so one can recognise intermediate tones between the metals. The seven fundamental metals are a harmony that was created by the pure planetary influences on the Earth. Subsequently the planets, between themselves, entered into relationships, aspects, (trine, square, opposition, conjunction, etc.). These accordingly produced new variations of joint planetary influences and through these arose the other metals.

It is possible to determine the relationship of these other metals with the pure planetary metals by the way in which they were and are linked with certain organic processes.

When one considers metals as we have done as something born out of light and air, one can comprehend how colour as 'the deeds and sufferings of light' * arose in the world.

*

During the last decades ever new and surprising discoveries have been made concerning the connection of metals with living substances and organic processes. In the course of time it has emerged that among the heavy metals there is hardly one that is not found in small quantities in coal and oil besides playing a decisive role in the life of plants, animals and man.

We will select a few examples. Firstly there are the metals related to iron. There is for instance manganese which is very widely distributed and appears in small quantities in the oldest rocks. In the schist and slate formations it forms ore deposits as does iron. The iron-magnesium silicates, which we have described as rock-forming minerals in chapter 13, are particularly rich in manganese. When one considers the life processes underlying these slates and greenstones as belonging really to plant life, then it seems that in the past manganese must have been a sort of *iron for the plant world*. Manganese has a peculiarity worth noting : in its natural state it is practically never compounded with sulphur (as iron is) but almost exclusively with oxygen. It also combines with silica as rhodonite and with carbon dioxide as rhodochrosite.

* Goethe.

The relationship that a metal has to other substances in nature gives some indication of whether it was connected with life, and therefore can connect with it again, or whether it is something cast off from the processes of the mineral-plant. One can nearly always say of metal ores which are combined with oxygen that they are cast off from vegetative processes. On the other hand the sulphur ores indicate an all-embracing life process and have a refining effect preparing the ground for higher forms of life. We already mentioned this in connection with iron pyrite, etc. The silicates and carbonates of metals also express a more plant-like nature just as the oxides do.

No wonder then that the organic remains of the Earth's past – asphalt, oil and coal – are rich in manganese. The massive deposits of manganese are to be found, as we have said, in the slates and younger strata, deposited while the old life died away more and more. We find manganese as pyrolusite and psilometane always associated with limonite (iron oxide). These are oxidic ores.

Manganese is necessary for many plants today. Manganese deficiency causes diseases in certain plants and prevents the flowering of tomatoes.

All living organisms form specific enzymes and similar substances which play a mediating role in the processes of digestion and breathing. In these there is always a complicated protein base combined with a metal. They are true metal-protein compounds whose metal content develops quite astonishing functional properties in connection with the transformation of other substances. Here we meet undoubtedly that former life condition when *everything mineral was embedded in protein*. Later individualised forms of life have retained something of this as the basis of their most vital life processes.

It would be quite wrong to imagine that a metal-protein compound possesses some inorganic chemical properties whose interaction produces, so to speak, 'life functions'. This is not the case. What happens is that the organic protein acts as a sort of 'window' for the forces that stream to the Earth from the surrounding stars. Through the metals being in suspension in a fluid state (a colloidal sol) in the protein, they become sensitive. What happens can be compared with the experiments of L. Kolisko. The metal embedded in living protein again becomes sensitive to the influence of the planet from which it originated. In this way, by means of metals, the living protein is made pervious to the in-raying forces from the surrounding heavenly bodies. The starry heavens influence life. But also the life and soul forces of the individual organism can work on its own bodily functions through these embedded metals.

When one considers metals as originally part of life and indispensable to life, one begins to find solutions to many problems that are difficult to explain from a purely scientific point of view. There is a small number of metals present in the Earth, more plentiful than for example antimony, uranium, silver, bismuth, quicksilver, gold and platinum, which are nevertheless not found in large deposits of ore but are so finely distributed that it is not possible to extract them in any quantity. They were discovered by spectroscopic analysis. One of these is scandium, a rare earth of which one finds infinitesimal traces in nearly all minerals which constitute stanniferous granite. It is also in all clay minerals and is easily detected by the spectroscope. Scandium was therfore already present in very early times of the Earth when granite began to form out of the mineral-plant. As we have said before, this mineral-plant life was of a low order and relatively undifferentiated. In a certain way it could be compared with our smallest present-day soil fungi – a form of life that has not developed further but remained stationary. Scandium was obviously important for these low forms of life, and indeed it has been found when studying the biological properties of scandium that simple moulds like aspergillus cannot live and grow without it. When one considers how such moulds seem to grow everywhere, one can appreciate the wide-spread distribution of scandium.

The metal gallium is another example. It is found in zincblende and grey copper ore (tetrahedrite) in sufficient quantities to be extractable. All clay minerals contain this metal. Its properties, like those of scandium, point to a relationship with tin, zinc and aluminium. Since aluminium silicates were the most important substance in the life of the old mineral-plant, it is understandable that gallium is vital for the existence of all plants today. If one carefully excludes gallium from growing plants the yield is reduced by up to 40%.

Another metal that has come to the fore with the development of radio transistor technique is germanium. It is found in tin ores and copper deposits, but the largest quantities of this valuable metal are obtained from coal. It must have played an important role in the fleeting plant forms and life processes of Carboniferous times. It is sufficiently abundant in coal to be extractable from the ashes.

Even in the old coals that precede the true Carboniferous period one often finds unusual concentrations of metals which show that these must have played an important part in the life processes that produced the coals. Thus we have, associated with Canadian pegmatites, a dense bituminous coal known as thucholite which contains not inconsiderable amounts of radioactive uranium and thorium. Likewise there is a very old Swedish coal which, besides radioactive uranium, also contains metals such as nickel, copper, zinc, molibdenum, vanadium, lead, tin and bismuth in the ash. These old coals are sometimes found in granitic rocks. They belong therefore to the time when life was not yet differentiated. If one considers that even today the lower plants such as algae can store and 'digest' uranium and radium, it is understandable that in early times quite other quantities of metallic substances were enveloped by life. Earlier it was assumed that this metallic and mineral content of coal ash was carried by water into the coal at a later date. It appeared unusual that such quantities should be present, for it was not known that metals were necessary for life. Meanwhile it has been found that these metals are also present in oil bound up with complex organic materials. Metals such as nickel, vanadium, copper and molibdenum are found in oil compounded with porphyrines. The composition of the porphyrines resembles that of the red haemoglobin of human and animal blood and is also closely related to the green chlorophyll in plants. So one sees that in these early times metals were connected with life to a far greater extent than now. The durability and persistence of the metal-containing oil porphyrines is such that they have endured unchanged to this day and are not even affected by the high temperatures to which they are subjected in the processes of refining. They can still be found in petrols and lubricating oils where they can have detrimental effects.

When we consider that vanadium and molybdenum are necessary

for the nitrogen bacteria on the roots of leguminous plants such as beans, peas, clover, etc., tungsten for other bacteria, etc., we can appreciate how vital is the role of metals even today though no longer in the quantities of former times.

Another metal, quantitatively next to iron, is titanium. The titanium content in rocks is about one tenth of that of iron. It occurs usually combined with iron as ilmenite ore but can also be found independent of iron associated with quartz as the beautiful rutilated quartz (Venus' hairstone). This shows that this metal occupies a position between iron and silica. Ilmenite (titanium iron ore), rutile and other titanium minerals usually occur in the oldest rocks which contain much silica. They are usually finely distributed and do not form ore deposits. Thus titanium can only be obtained when after weathering and transportation, its heavy particles sink to the bottom as ilmenite sand. These dark sands are fairly widespread on many coasts. Through its presence in the oldest silicic rocks titanium shows its connection with the life of the mineral-plant and even today nearly all plant leaves contain titanium.

Chrome is also found with iron as chromic iron ore (chromite). It forms few ore deposits and mainly occurs finely dispersed in serpentine and other greenstones. It gives both rubies and emeralds their colours. Chrome is a hard metal and we find it in the densest and toughest rocks belonging to what we have called the tree stage of the Earth.

All these metals (which are related to iron) including vanadium, molybdenum, tungsten, cobalt and nickel are found in the oldest strata. They are present in perceptible quantities, combined with proteins, in the body fluids and organs of living organisms where they are known as trace elements. One can perceive the effect of minute quantities, but these effects cannot be explained purely chemically. The significance of these effects is due to their dilution and distribution i.e. their enormously increased surface area. Forces then come into operation which have nothing to do with ordinary chemistry.

If one goes one step further one reaches a domain not known by modern science and one comes to the effects of those metals we have not yet dealt with i.e. lead, tin, gold, silver, quicksilver, etc. These metals are found, like iron and copper, in fairly plentiful deposits. Only gold tends to be finely dispersed. To understand the role of these metals in the life of the Earth one should know something of their planetary origin and their relation to the inner organs of higher forms of life. Fundamentally the planetary metals should be considered as *organs of the Earth* and the large ore deposits in the Earth recognised as the dead remains of these organs. Once upon a time they were living, in-streaming activity of spirit-filled creative powers from the surrounding cosmos which then coagulated to dense earthly matter. Wherever we find large quantities of these metal ores concentrated and embedded in the rock, we must picture in former times mighty planetary forces streaming in from outer spheres. Thus, in coloured clouds, the metals came to the Earth. Later they condensed to the liquid state and flowed into the fissures of the Earth. This explains why all large metal deposits peter out as one follows them deeper into the earth. It is erroneous to think that metals came up from the interior of the Earth. They came from outside as meteoric iron does to this day.

It is a mistake to draw conclusions about the constitution of the Earth 6000 km down when one only knows a few kilometres of the crust.

The sparse distribution of metals in the deep-seated rocks and the sudden petering out of nearly all sizeable ore deposits at depth suggests quite another picture. Originally everything metallic was in an airy state around the Earth. As it streamed in, in the early times of the Earth's life, it became life process and was finely distributed in the rocks. Only later as life began to differentiate – in the schist and porphyry period – did the metals form ore deposits. This is why the major deposits coincide with the transition from the primal rocks to the slates, schists, greenstones and porphyries. The 'organs' of the Earth were first formed in the fluid-living state. Then lead acted on the inner framework of the Earth, tin created the correct balance between solidity and fluidity, gold with its urge to disperse held sway between gravity and levity in the living strata of the old atmosphere, silver regulated heat and mercury the circulation of fluids in the living Earth.

All living organisms have retained from the past something from the living influence of metals which extends beyond mere crude substantiality. To understand the 'life of metals' in the creation of the Earth, one must seek their action in man, animal and plant. Here they are still as active as they once were in the Earth as a whole. Recognising the effect of trace elements is a step in the right direction. These 'traces' however conceal even greater secrets which lie behind the crude mineral matter in the creation of man, animal and plant. Therefore a future science will attempt to solve the riddles of the mineral world by studying the embryological development of living organisms, especially man.

15 *Embryonic Development and the Basic Processes of Rock Formation*

The innermost purpose of creation from its earliest beginnings was the development and emergence of man. What we see around us as the mineral, plant and animal kingdoms are *not* the ancestors of present day man, but are the cast-off husks of an all-encompassing being that had to lay aside its sheath in order to free itself from the compulsions of its own evolution. Nature arose in order to give a spiritual being the possibility to rise from its dependence on matter and seek in freedom its spiritual origin. Yet out of the creation of the world as a *totality* there emerges man who, in so far as he is a being of nature, is able to *forget* his spiritual origin. Since he is aware of his natural self, he is placed in freedom to remain a creature of nature *or* to progress towards further developing his spiritual self. If man does the former, he makes what should be a stage of development, into a fixed condition and separates himself completely from the evolving world. This leads to hardening, death and self-destruction, as the present state of humanity suggests only too clearly. If however he recognises himself as a spiritual being and overcomes and transforms his natural state, he returns to his source as a *free* spiritual being to take his place among the creative and progressing powers of the universe.

The physical body of man submits to the inexorable laws of birth and death, growth and nutrition. This body is permeated with solid mineral substances, fluids, air and also heat. As we have seen, these four elements are also the stages of the coming into being of the Earth. The beginning is not the creation of visible, mineral matter, but is *heat*. In the primal beginning of creation the human spirit was embedded in warmth, his body was flowing streams of warmth. This body had neither life nor feeling, nor did it offer the possibility of developing a consciousness of *self*hood. The present day mineral is at this stage of consciousness. The whole world of rocks has remained at this stage, left behind, one might say. Rudolf Steiner has called this first evolutionary stage of Earth and Man 'Ancient Saturn'.

The next stage, called the 'Old Sun', brought life to this body of warmth, and with it the formation of air and light. Man became a plant-like being with a body of warmth and air. The whole plant world of today is at this stage. Thus a part of what was originally created moved on a step, and part remained behind. This last development was preceded by an abbreviated recapitulation of the earlier stage in a transformed condition.

At the third stage, called the 'Old Moon', the human 'body' was endowed with *feeling* and now consisted of warmth, air and water. Besides the fluid element, sound came into being in this 'Moon-Earth'. Man advanced to an animal stage. What remained behind at this stage later formed the animal kingdom.

As the human being entered upon the *real Earth* evolution, he first had to recapitulate the previous stages again in order to incorporate solid mineral matter into the physical visible form of today which has become the foundation of his awareness of self.

In these recapitulations we have the familiar biogenetic law propounded by Ernst Haeckel which states that the organic development of an individual organism is a contracted recapitulation of the history of its species. The history of the human species is not the development from bacteria to man, but from a spiritual seed surrounded by warmth to the present mineral-filled visible form of the human body, able to be the bearer of a spiritual being conscious of itself. His *body* is the oldest and most perfected possession of man. When the human individuality prepares to re-enter earthly existence, it needs the help of all the creative powers which have taken part in the creation of the human body from the beginning. In the building of this body, sheathed in the mother's womb, the whole evolution of man and Earth is recapitulated.

We can follow the processes of the 'real Earth' evolution in the embryological development of man, recognising three stages. The first is the morula stage where, from the fertilized unicellular egg, a globular structure somewhat resembling a mulberry (morula) is formed. It has neither inside nor outside : one might describe it as *granular* – formed out of many equal cells. The second stage is that of the flat surface of the germ layer stage. With this begins the differentiation of organs. By invagination and upward surging the cavities of the digestive tract and the spinal canal are formed. The third stage begins with the depositing of calcium in the cartilaginous skeletal system, in other words – ossification.

If one surveys these three stages of early embryonic development up to the onset of ossification of the skelton, one has, as regards structure and form, a faithful picture of the basic structure of the world of rocks as we described it. There is the granular structure of the primal rocks, especially the granites, which is followed by the flat layering of the schists and slates, which are immediately acted upon by the cavity-forming processes of porphyry; then calcium gradually infiltrates, becoming predominant in the Jurassic period.

It is important to realise that in this parallel, human evolution is the primary phenomenon, and events in the world of rocks are the results. This fact shows the rocks from a new point of view. They have an intimate relation to the development of the human form. They are the cast-off husks and residues from the coming into being of this form.

Man however is not only a form, he also has life, feeling and conscious thought. One can ask: where are the cast-off husks from the development of life and feeling to be found? We have to remember that man during his development also left behind the plant and animal kingdoms. As the rock kingdom is related to the development of his *form,* so is the plant kingdom to the development of his *life* and the animal kingdom to the development of his soul and *feelings*. In the realm of individual conscious thought, however, man rises above the kingdoms of nature to what is specifically human.

To return to embryology and rocks – we spoke of three processes: morula stage, germ layer stage (differentiation of organs) and the beginning of calcification. Threefoldness is also to be found within the second stage. Three different layers develop – the outer germ layer or ectoderm, the middle layer (mesoderm and mesenchyme) and the inner layer or endoderm.

Each of these layers takes part in the forming of certain regions and organs of the human body. The ectoderm is the origin of the greater part of the sense organs, the skin, nerve cells, and also the lenses and vitreous humour of the eyes. All these have a special relationship with the silica processes in rocks, particularly evident in granite. Considerable parts of these organs contain silica or else silica processes play an important role in their functions.

The middle layer is mainly concerned with the creation of the organs of breathing and blood circulation – heart, blood vessels and blood cells. Apart from the lenses and the vitreous humour of the eyes, it takes part in the building of the predominantly skin-like surfaces of all organs. The membranous creations of the mesoderm are the schist/slate process within man. The mesenchyme part of this middle layer is the builder of the supporting and filling-in tissues between what is formed by the other layers; hence it is also responsible for the basic tissue of the bones.

The calcification of the bones originates however from the inner layer (endoderm) which forms the nutrition system of the developing organism. The membranous lining of the whole digestive tract proceeds from the endoderm.

In the mesenchyme part of the middle layer lies the origin of the important process of cavity formation; it also helps to give structural and functional details to the various organs. All organs embodying cavities, including the bones, proceed from the mesenchyme. We have the counterpart of this embryonic process in the porphyry formations of the rocks.

Through the activity of the entire middle layer i.e. forming surfaces and cavities, there develops a system of organs that is a reflection of the interplay of the surrounding planetary system. Man, in his development, has taken into himself all that is around him. Not only does he partake of the lower kingdoms of nature but also of the surrounding world of stars. We can only touch on this here. Rudolf Steiner has dealt with it extensively in his writings and lectures.

The inner organs of man correspond in a certain way to what the metals are in the Earth, which, as we have shown, originate from the planetary spheres. Just as within man these organs originate from the middle layer (corresponding to the schist/slate and porphyry processes in rocks), so we find the main deposits of the seven planetary metals in the schistose and porphyritic rocks of the Earth. One would never look for substantial metal deposits in the typical granites nor in the Jurassic limestones. It is where granite begins to give way to gneiss and schist or in the early porphyries (pegmatites) that the metals first appear in seams, deposits and veins. In the real primal granites the metals are extremely finely dispersed, as we have seen.

In the true limestone formations, the metals again recede and we only find them when the limestone is interlayered with slates, sandstones or volcanic rocks. In most cases however, these younger ore deposits are of a secondary nature i.e. they have arisen through the disintegration, dissolution and re-deposition of older metal-bearing strata.

With these younger limestone deposits we touch on a process outside man which in man is connected with the origin of the digestive organs during the germ layer stage of the embryo. We have seen that the membrane lining these organs originates from the inner layer or endoderm. This applies especially to the alimentary canal and its glands, the inner tissues of the liver and pancreas, the inner lining of the lungs and part of the organs of taste. In the forming and functioning of these organs, the nourishing and building-up forces of calcium are active, but they do not here reach the point of forming and depositing the substance calcium. If one brings to mind the origin of the limestone mountains, one can understand that calcium, on the one hand, harbours nutritive and up-building forces, and on the other hand, gives man his inner mineral skeletal frame. The connection of calcium with the animal world is evident when we consider that the major deposits of it in the Jurassic and Chalk formations are due to vast quantities of shells of small organisms arising from the tremendous flooding life of the Earth's past. Such abundance of *formed* life is only possible through the excessive nutritive and up-building forces of calcium. The older life of the Earth with its mineral-plant and plant-animal, predominantly dependent on silica, was not like this and left no concrete forms behind – only impressions of attempted forms. From the silica processes of the past have developed the formative forces working on the periphery of the human body, while the inner structure is due to calcium processes. It is of particular interest to note how calcium is transformed when it enters the skeletal structure. The calcium which forms our bones is not calcium carbonate like the limestone mountains, but mainly calcium phosphate. Calcium carbonate has more of a connection with vegetative life processes while calcium phosphate provides evidence of the active nutritional process of the animal organism that took hold of the calcium. Phosphorus plays a most important part in our whole metabolism. This is noted in passing and cannot be elaborated further at this point.

The ossification which begins from the nutritional function of the endoderm does not proceed uninterruptedly. Regarding the very complex process of ossification of the skeleton certain facts have recently been established which support the principle of recapitulation. It has been found that in the embryo this process is initiated four times, and three times it retreats completely. Ossification begins thrice from different classes of tissue (cartilage or connective tissues), and before each new beginning the calcium is completely withdrawn and only at the fourth onset does it proceed uninterruptedly to post-natal time.

This fact – incomprehensible from a utilitarian point of view – can be explained if one sees, in this four-fold repetition, a picture of the whole creation of man and the Earth. What took place in the stages of human creation, from the embedding of the human spirit-seed in warmth, to his body woven of air and life, to his feeling, fluid body, up to his present body permeated with mineral substance which is the basis of his present consciousness and power of thought – all this is reflected in embryonic development.

*

This approach to the study of geology based on the development of man – only briefly outlined here – is not only the most important but also the most difficult. Its importance will be appreciated if, in the future, the animal and plant kingdoms are thought of as linked with human development. In the combining of natural scientific data with the results of spiritual scientific investigation lies the universality towards which so many strive. Creation itself is a unity, it is directed towards one goal – mankind.

16 *'Time' in the Development of the Earth*

In the previous chapters nothing has been said about a factor that is of great importance in modern geology, namely – time.

Human thought has derived its measure of time from the circling of the Earth round the Sun, of the Moon round the Earth and the course of the Sun through the zodiac. The fact that the regular repetitions of these motions, through thousands of years back to the earliest known civilisations, tally with our present-day calculations, has led, with some jusitification, to the idea that these familiar rhythms of time – years, months, weeks, days and hours – hold good for millions of years before and after the present time.

However, besides this perfectly logical conclusion, there is another equally valid, namely, that all repetitive occurrences point to something living, developing, ever-renewing itself. In contrast, the mineral kingdom – crystals, ores and rocks – which is cast off from life, presupposes duration – a tendency to persist in its form. This persistence is to some extent an illusion, for, as we have seen, in weathering and dissolution mineral material can be led back into the stream of life. On closer investigation, as we shall endeavour to show towards the end of this chapter, this 'duration' of the mineral is part of a far greater rhythm than is usually imagined. There is nothing really dead in the world, only varying cycles of becoming, existing and passing away.

Nonetheless we can say as regards the present age: Life exists in 'time' and what falls out of life exists in 'duration'. Strangely enough though, even in the rhythms of our time there is an element of duration in so far as the recurring of these rhythms is calculable. Yet ultimately, all rhythms of anything that is alive are incalculable.

From the fact that what we call 'time' – determined by the relation of the Earth to the surrounding stars – has a certain periodicity, it follows that all the participants in the rhythmic phenomena stand in a living relationship to each other, i.e. they must be themselves living entities.

All living beings are the result of a development from the past into the future, and have, in the early stages of their existence, different rhythms of growth from those in middle and later life. One has but to compare the pulse and breathing of a baby with that of an adult and of an old person.

As we have seen the Earth has passed through stages of a development arising from creative deeds of spiritual beings. These stages took their course in a cosmos which did not have the settled state of our present world. To enable the present order of our solar system to arise, one by one the planets and finally the Sun and Moon had to separate from the developing Earth (which included man and the emerging kingdoms of nature). Thus, up to the separation of the Moon, 'time' could not be reckoned in our present years since the Moon as a heavenly body was not yet incorporated in the system. Only after this event could our planetary system begin to assume its present order and rhythms including its relation to the zodiac. Between the separation of the Moon and the attainment of the present calculable periods of revolutions, time periods approximated more and more to our present years; but they are nevertheless not to be reckoned in exact years.

In this connection Rudolf Steiner repeatedly observed that vital turning points in evolution, such as when 'time' in the modern sense of calculable rhythms began, are always 'connected with certain positions of the heavenly bodies' (lecture, 31.12.1911)*. Naturally such constellations could only occur after the Sun had separated from the Earth and begun its course through the zodiac. A predictable and exact path was however not achieved immediately after the separation of the Sun. One can say it took thousands and thousands of years (though not measured in years, since they did not exist) till day and night periods, and therewith approximate yearly rhythms, harmonised. To complete the organic process towards calculable 'time' the Moon had to separate from the Earth and start on its own orbit, exerting its influence from outside.

For those interested in the astronomical aspect we recommend E. Vreede's 'Anthroposophie und Astronomie' (Freiburg/Br. 1954).† This work, which elaborates Rudolf Steiner's findings, deals with the

**The World of the Senses and the World of the Spirit.*

†Translated as *Astronomical letters 1928–30* and privately duplicated.

precession of the equinoxes, aphelion and perihelion, variations in the eccentricity of the Earth's orbit, obliquity of the ecliptic (angle of the Earth's axis) etc. From this it appears that in the 20th millenium B.C. the 'time' began when it became possible to reckon time in years in a universe now moving according to mechanical laws. This point in time is approximately the middle of the Atlantean epoch and therefore of the Earth evolution. Some 4000 years later i.e. in the 16th millenium B.C. the Ice Age began. This period of the Atlantean development was one of extraordinary climatic variations alternating between widespread glaciation and almost subtropical conditions. What occurred is not only due to the constellations (planetary aspects) we spoke of but also connected with the development of humanity. It is only a few thousand years since man walked the Earth in his present form. The prehominids which have been found in all continents were creatures whose forms hardened too early, but they have nothing to do with that part of humanity which held back the solidifying of their bodies until they could become apt vehicles for souls gifted with self-awareness.

The climatic alternations were the instruments of creation either to school man or to wipe him out if too weak.

During the Ice Age not only were the climatic zones reorganised but the atmosphere also underwent a complete change. In his book *Cosmic Memory* Rudolf Steiner describes how the Atlantean atmosphere contained far more water than today – the air was 'thicker' and the water 'thinner' resulting in a completely different interplay of the two elements. This produced the misty land of North Atlantis, the climate of which was decisive for the development of the white races. In nordic mythology North Atlantis appears as 'Niflheim' (home of mist) and the warmer Southern Atlantis as 'Muspelheim'. The misty land of Atlantis did not know the rainbow because the Sun could not penetrate the thick mist. Not until the Flood of the 10th millenium B.C. and the end of the Ice Age did water in masses pour down from the atmosphere, the sky clear and the first rainbow appear as recorded in the Bible.

From this it follows that there are certain limits to the validity of our calculations of time. When one observes natural phenomena such as the depositing of mud and sand in lakes, river estuaries or bays – or even radioactive decay – it is most important to be aware of the limits of the duration of our time scale if one is not to fall into the error of projecting the laws of the now aging Earth into the time when the young Earth was coming into being.

It would be as if a scientist, after noting the pulse and respiratory rhythm of an adult throughout one year, would proceed to make a graph and projecting the curve deduce what his pulse and respiration would have been 300 years previously and would be 300 years hence. The observations of the scientist are unquestionably accurate and his calculations equally so, but the pulse and breathing of the man do not continue for 600 years.

It is the same thing as when astronomers, basing their calculations on present-day slight changes in the precession of the equinoxes, the angle of the Earth's axis, aphelion and perihelion, which undoubtedly hold good for 25,000 to 30,000 years, project these rhythms without limits into the past and future, and therefore reach false conclusions.

In this way calculations have been made which go back 900,000 years and produce a curve of the intensity of solar radiation. In the geological field it was possible to trace a corresponding series of glacial deposits for part of this time, which showed a remarkable parallelism. The parallelism is not questioned; only whether it can be measured in years.

Another example is reckoning the age of the Earth from the present-day rate of decay of radioactive elements. These calculations are based on the supposition that the decay of radioactive elements 'began' simultaneously with the formation of these elements in the cosmic process of solidification. Therefore, since the quantities of the end products of decay in a given time are calculable, one can determine how 'old' a particular mineral is and thus how old the whole Earth must be. This results in ages varying from 200 million years for a uranium pitchblende from Joachimstal in Bohemia to 2,600 millions for a monazite from a Rhodesian pegmatite. Equally tremendous differences in the rate of decay of radioactive elements in a given time occur when calculations are based on contemporary phenomena. In this way one obtains the radioactive half-life i.e. the time taken for the number of atoms of a particular substance to be reduced by half. These times range from millions of years in the case of uranium and thorium to fractions of a second for radium C, a decomposition product of radium.

An extremely interesting phenomenon in connection with rates of decomposition is that the shorter the life of the radioactive element, the stronger the penetration of the alpha particles. Intensity of radiation stands in an unmistakable relation to time. The questions arise: why is the decay of different radioactive elements so variable in time, why do atoms only decay by stages as they give off energy, and why does this decay not occur all at once? There are no simple answers to these questions as long as one considers *naturally* occurring radioactivity only.

Strange to say, man has been able to control the time factor of radioactive decay and regulate or check the natural process. He has also been able to produce artificial radiation electro-magnetically. Basically the processes have been brought under control by embedding large quantities of small units of pure radioactive substances, such as uranium, in graphite or heavy water.

By contrast, the changes, which take place very slowly in nature owing to the fine distribution of these substances embedded in the rocks, can be tremendously accelerated and enhanced by bringing together large quantities of radioactive material leading to the catastrophic chain reaction.

Thus on the one hand we have the principle of the atomic reactor producing energy from controlled decay and on the other hand the atomic bomb.

The intensity of decay depends on the concentration of the mass. The extremely slow, natural process of transformation of uranium 238 into plutonium 239 can be either directed to the production of energy in an optimum of time within the atomic reactor, or it can be made to take place almost instantaneously in a fraction of a second in the atomic bomb.

A small quantity such as 200 grams of uranium 235 or plutonium 239 is harmless, but if it is increased to several kilograms (about the size of a coconut) the chain reaction sets in which leads to detonation. In practice this reaction is triggered off by bringing together two hemispheres of uranium 235 or plutonium 239. Thus it is clear that there exists a relation between the *weight* of a pure radioactive substance and the *time* it takes to decay.

Since pure substances do not occur in nature, the intensification which can be obtained only from pure uranium or plutonium will never take place. For rapid disintegration a relatively pure radioactive substance is necessary, so that little of the radiation connected with the formation of neutrons may be diverted by being absorbed into foreign stable material. Since in nature these substances do not occur in a pure state, such violent reactions will never take place.

It is conceivable however that natural radioactivity has not always existed at the same intensity as today. The rate of decay of certain elements may have increased during the passage of time and not maintained a steady course from time immemorial. This may appear to be incompatible with the presence of lead and helium in radioactive minerals since these are presumed to be the stable end-products of decay as observable today. This seems to be borne out by the different atomic weight of this lead compared with lead ore found far removed from any radioactivity. It could however equally well be that the ubiquitous helium and the lead were always present in these minerals. Thus it is possible that radioactivity is not of the great age generally surmised. Errors like this arise from projecting present-day phenomena into the past and the future.

17 *The Earth as Seed of a New World*

The solid element of our planet Earth – the rocks – appears to us as a realm apparently completely at rest and no longer in the process of becoming. The rocky summits of mountains have remained almost unchanged for thousands of years, although in further millenia weathering will reduce the created world to dust. The weathering is due to the elements from which life unfolds – water, air and heat. Life as revealed in the higher kingdoms of nature (plants, animals and man) has nothing of the quiet and settled state of the rocks. It pulsates in the plant between seed and fruit, in the animal between birth and death and leads man beyond birth and death to find his way to his spiritual origin at a higher stage of existence.

The time during which the rocks arose came to an end thousands of years ago. Today the stage of life of the Earth is comparable to a ripe seed which has fallen to the ground and rests in the winter soil awaiting germination. This present stage is best expressed in such imagery taken from the plant world, for the plant seed is a completed stage which can lie dormant for shorter or longer periods. It is also something hardened and mineralised and contains only a minimum of life. Ripening is a mineralising microcosmic process which corresponds to the macrocosmic process of crystallising and hardening of the world of rocks. The hardening of the seed, like the hardening of the rocks, is the unavoidable pre-requisite for the emergence of a new plant or of a 'new Earth'.

What takes place when a seed germinates? Water, air and warmth act upon the hardened seed breaking down its substances into non-organic substances. Only after this dissolution and reducing to chaos, can the constructive and revivifying formative forces of the plant take hold and call forth the seedling.

In one of his last 'Letters to Members'* in January 1925, Rudolf

*Published as *The Michael Mystery* 1956

Steiner posed the question: *What is the Earth, in reality, in the macrocosm?* His answer is that spiritual research reveals that the original state of the macrocosm was extraordinarily full of life which gradually died. The macrocosm became more and more subject to calculable laws. The life that has died out of the macrocosm has become microcosmic and is to be found everywhere where seeds germinate and embryos develop in nature.

The present Earth, with its kingdoms of nature and man, developed from the process of dying. The dying of the macrocosm was the necessary condition for the development of man as a being conscious of egohood. Through these processes man and all nature carry something germinal into the future.

At the end of the above letter to members Rudolf Steiner calls the Earth 'the embryonic beginning of a new-arising cosmos.' In the microcosmic life of the kingdoms of nature there is today a super-abundance of seeds and eggs which perish and one may well ask what becomes of these lost forces. According to Rudolf Steiner the superfluous seed forces of the plant world provide the *substance* for a new image of an ordered macrocosm. The superfluous forces of the mineral world carry the forces from the plants to their allotted places in the macrocosm. The super-abundant germinal forces in the animal world act in such a way that all that is carried out from the plant world by the mineral forces into the universe is held together in a sphere (a globe) so as to present the aspect of a macrocosm rounded in on all sides and self-contained.

Here Rudolf Steiner points unequivocably to germinal forces in the mineral kingdom. Where are we to find these if we look deeper into nature in the light of spiritual cosmology? We have seen that all the rocks of the Earth initially arose from tremendous plant and animal life-processes. These were predominantly life-processes of the mineral-plant and the plant animal. This can be recognised by the fact that over 90% of the rock mass is composed of silica minerals and only about 0.2% consists of calcium i.e. animal residues. Thus rocks are predominantly formed from vegetative life-processes. The young limestone strata are but a thin skin or shell patchily covering the Earth.

Everything resulting from dying vegetation contributes to a new life. This applies not only to seeds but also to the fallen leaves and in fact to the whole plant as it withers and dies. The seed can become a new plant or it can, together with the rest of the plant, rot and be absorbed into the living humus of the soil. Under natural conditions the dying plant always contributes something towards the maintenance of the life in the soil. The germinal nature of this decaying process is not obvious and is only revealed in the formation of humus when the plant residues are united with the minerals of the earth. At first the plant tends to dry out and harden and then it disintegrates and provides the material for the activities of lower organisms. The resulting substances resemble inorganic matter without actually becoming mineral. These processes resemble fundamentally what happens when a seed is formed and, after the winter's rest, proceeds to germinate. As regards the substances present, there are of course great differences, but the *processes* are closely related.

When one transposes these ripening and dying processes of the plant to the macrocosmic scale of the whole Earth one approaches the idea of the birth of the rocks. Silica then played a similar role to carbon in the plant today.

The process of rock formation came to an end about the middle of Atlantis – some 15,000 years ago according to Rudolf Steiner (lectures to the workmen 17.2.1923). Rocks at that time were not so hardened as they are today. The hardening process continued for many thousands of years. Up to the last millenium B.C. enormous buildings were erected by many different peoples using crystalline rocks which today are excessively hard. None of these peoples were acquainted with iron or steel, yet the blocks of stone were dressed with such precision that one cannot insert a knife blade into the mortarless joints. This suggests that the rocks were not so hard in those days.

Even today it is known that freshly-mined granite, sandstone, slate and limestone are softer and can be more accurately dressed than the same rocks after exposure to the air for some time. Air and warmth dry out the rock and often cause chemical changes. One of the most striking examples of this hardening process today is the emerald. In some localities these gemstones are found embedded in soft mica-schists. If one extracts them one can find that – in spite of their great beauty and transparency – they are so soft that one can crush them to a damp powder between the fingers. These soft crystals must be handled very carefully and must be kept for some weeks in a closed wooden box to dry out slowly. They then become as hard as rock crystal. This is why most emeralds show a network of fine haircracks.

The hardening of rocks is not a continuous process. It is alternating and cumulative. In one of his lectures to the workmen (20.9.1922) Rudolf Steiner said that the hardening began after the separation of the Moon in Lemurian times. In certain areas the Earth solidified to the density of a horse's hoof and after a while became softer again. The hardening then began again in another area and so on. In a later lecture (17.2.23) he described this rhythmic process from a different aspect, showing how the condition of the Earth is dependent on the course of the Sun through the zodiac. Thus 25,920 years ago (one platonic year) the Earth was in a somewhat similar condition to that of the rocks today. Then as now the Sun stood in the sign of Pisces at the vernal equinox. This coincided with the end of Lemuria. The alternating processes of hardening and softening which set in after the separation of the Moon had, by the end of the Lemurian time, led to a condition which can be compared with that of today. In the same lecture of 17.2.1923, we are told that in between, when the Sun (at the vernal equinox) stood in Libra, the Earth became again soft and pliable and was like a living plant. That was the time, about 15,000 years ago, which we mentioned as the end of the time when new rock arose. From this time on began the hardening which has reached its peak in our time; in fact has already passed it. At the end of the current platonic world year (extending from the vernal equinox in Libra about 15,000 years ago to its return to Libra in about 11,000 years) the Earth will again be in a living plant-like condition.

From all this it becomes clear that the present mineral state of the rocks only exists within one platonic world year. At the beginning of this world year when the vernal equinox was in Libra there was a balance between the fluid and solid states of the rocks. The hardening process has not completely penetrated the interior of the Earth even now, as we have seen by the effect of air on rocks when brought to the surface.

Before the rocks became so hard and dry that the elements of life – warmth, air and water – could no longer affect them, disintegration set in through weathering and other destructive forces. Natural radio activity is of course one of these – the self-disintegration of the densest materials of the Earth. Radioactivity has even the power to reverse crystallisation and transform crystalline materials into gel-like substance (see Chapter 11). We are living on an Earth that is already on the way to the time when the vernal equinox will again be in Libra and the Earth will be in a state of balance as it was 15,000 years ago.

Mineral existence is therefore only a transitional stage. It is not a state of complete death but is a prolonged stage of subdued life which once again will awaken to full force, but transformed. We must now return to our comparison with the dormant plant seed.

In the letter to the members previously quoted, Rudolf Steiner has emphasised the significance of the surplus germinating power in the kingdoms of nature. What flows out from the plant world provides, in a sense, the material for a new-arising picture of the macrocosm i.e. the future Earth. What comes from the plant will be conveyed to the appropriate place and moulded by the surplus forces of the minerals. And what flows out as surplus reproductive forces from the animal kingdom acts on what has gone out from the mineral and plant kingdoms and forms it to a sphere, rounding it and holding it together so that it presents the picture of a self-contained macrocosm. Rudolf Steiner pointed here to a macrocosmic law which throws light on many natural phenomena. It reveals the inner meaning of the interaction of the three kingdoms of nature. What is described as the working of the forces of these kingdoms of nature for the future of the macrocosm must also find its reflection in the microcosm. One may recall the old hermetic saying: 'As above so below'.

If one accepts this law one may well ask where it is manifested in the world of the smallest entities. Here the scientist's mind turns towards the manifold world of symbioses, commensalism and other intimate relationships which exist between the three kingdoms of nature – a world full of puzzles, whose laws can be stated but not explained.

There is a fundamental process in which the harmonious interaction of the three kingdoms of nature follows the macrocosmic law and can be seen reflected microcosmically. This is humus formation – the living soil in which plants can grow and which supports the higher forms of life. We have described elsewhere that this humus formation is a transformed process from the past (Chapter 4). Here it can be shown how, from another aspect, this process corresponds to the macrocosmic law.

We pointed out that everything proceeding from the dying plant bears something of a possibility of future life within it. One can consider it as surplus germinating power. All that withers and dies has the potential of taking part in germinating life. If one burns the plant remains, these potential forces unite with the picture of the new macrocosm. If, however, one composts them with soil as described by Rudolf Steiner in his Agricultural Course, the surplus forces of the plants are united with the forces proceeding from the disintegrating minerals. These forces somewhat resemble what takes place in germination. What finally results from the complete breakdown of mineral substance can be seen in the cloudy water of our streams and rivers after heavy rain or the melting of snow. The breakdown into fine particles also takes place continually in undisturbed soil through the excretions of plant roots. Something arises like a sort of mineral milk. In this substance new minerals are formed which however do not crystallise but remain in a state like curdled milk. This 'jelly' is the bearer of the above mentioned potential-germinative forces of the mineral. This strange substance would 'like' to become crystalline, but remains in this state and is to be found in the soil together with the humus from plant residues. It has a strange attraction for, and unites with, the organic part of humus.

When the minerals take hold of the plant substances in this way, a certain stability ensues which was not present in the individual substances while they were separate. Thus what comes from the plant 'is carried to its appointed place'.

There is however a third process. The 'certain stability' implies something incomplete, not yet finished. To bind this into a stable humus, the animal world must take part. The mineral-vegetable material is taken in and digested by certain lower animals. They are either true insect larvae or animals which remain throughout their lives at a larval (germinal) stage like the earthworm. In the digestive processes of these animals the mineral-vegetable substance is bound together and rounded off into a stable material by the co-operation of the germinative growth forces of the three kingdoms of nature, something quite new is formed which is the foundation of life for the whole Earth. Already here on Earth these forces work into the future.

If one observes these formative processes of the living soil and compares them with the picture of the macrocosmic process of which Rudolf Steiner speaks, it is not difficult to recognise the macrocosmic law which underlies the maintenance of life on the earth.

Thus even the mineral state appears as all-embracing embryonal force for the future just as much as the other kingdoms of nature. Man, however, is the 'Salt of the Earth'; he is the guardian of all the germinative forces of the Earth. It is given to him to work hand in hand with the creators of the world for the future of the Earth. From a spiritual knowledge of the microcosmos and the macrocosmos he will learn to treat the Earth as do the creative forces of plant growth, which every spring enkindle anew the microcosmic salt process in root formation from which the new plant can grow. Then, from man's husbanding of the Earth, a new Earth can arise in the future.

Table of Geological Formations

These tables, grouping the rocks into 12 sedimentary and 7 igneous formations, were made by Rudolf Steiner in 1890 when collaborating in the production of *Pierers Lexikon*, edited by J. Kürschner. It is reproduced from the *Literarischen Frühwerk* Vol. IV, Book 19, Dornach 1941.

(Additions by the author are in italics.)

The translators have omitted the names of a few minor local German series, unfamiliar to English-speaking readers.

A. SEDIMENTARY

Formation	Series	Fossils
V Anthropozoic or Recent		
12. Quarternary	b) Alluvium–marine and freshwater a) Diluvium–glacial and periglacial deposits	3 Large mammals including mammoth, cave bear, primitive man
IV Caenozoic		
11. Neogene (*Pliocene Miocene brown coal, limestone, salt, gypsum*)	c) Freshwater deposits b) Brackish water deposits a) Mediterranian marine deposits	2 Large mammals including mastodon, dinotherium, aceratherium, apes, giant lizards, palm, fig, elm, birch
10. Palaeogene (*Oligocene & Eocene basalt, oil*)	b) limestones, sandstones, clays and marls	1 Large mammals including palaeotherium, nummulites, fucoids

III Mesozoic		
9. Cretaceous (*oil*)	c) Sandstones, clays, limestones, Quader Sandstone, Cenomanian, Turonian and Senonian Chalk, Gosau formation	Last belemnites and ammonites
	b) limestones, sandstones, marls, gault, red marls	First deciduous trees
	a) limestones, sandstones, clays, marls	Sponges, foraminifera, spatangites, ammonites, belemnites, rudistids
8. Jurassic	d) Wälderton transitional beds	Giant saurians
	c) Malm or White Jura, Oxford	Reef corals, first bony fish (*birds*)
	b) Dogger or Brown Jura	Marsupials
	a) Lias or Black Jura	Belemnites, ammonites, sea saurians, cryptogams
7. Rhaetic. Dolomite Mountains	Dachstein limestone and dolomite, Kössener shales	Oldest mammal remains (Microlestes antiquus)
6. Triassic	c) Lettenkohle Keuper (*marls, sandstones and gypsum*)	Bipedal dinosaurs, *crocodiles*
	b) (*Muschelkalk*) dolomite, gypsum, salt	Sea lilies, ceratites, first crayfish
	a) Bunter sandstone, conglomerates, marls	Giant horsetails, conifers, first traces of birds

II Palaeozoic		
5. Permian	b) White and gray beds, Kupferschiefer	First amphibians heterocercal tailed ganoid fish
	a) Rotliegendes or totliegendes (*salt*)	Ferns, palms, conifers
4. Carboniferous	b) Productive coal measures (*oil*)	Terrestrial cryptogams, first conifers, spiders and insects
	a) Millstone Grit (*culm*)	Last trilobites, productus
	Moutain Limestone (*oil*)	

A. SEDIMENTARY—CONTINUED

3. Devonian (*oil*)	c) Shelly limestone, cyprid slates, Old Red Sandstone	Armoured fish
	b) Eifel limestone	Brachipods
	a) Graywackes	Primitive land plants, corals
2. Silurian	d) Upper silurian, limestones, slates	Seaweeds, corals, sea-lilies, brachiopods, first traces of fish
	c) Lower silurian, quartzites, shales (*oil*)	
	b) Ordovician, graywackes and slates.	Trilobites, brachiopods
	a) Cambrian, conglomerates, quartzites, shales	First unmistakable organic remains
I Pre-Cambrian	d) Primal slates	Eozoon canadense in early limestone, if organic, is the oldest fossil
	c) Mica schists	
	b) Laurentian gneisses	
	a) Bojische gneisses	

B. IGNEOUS

III Caenolithic	7 Younger trachytes and basalts
II Mesolithic	6 Younger porphyries
	5 Younger greenstones
	4 Younger granites
I Palaeolithic	3 Older porphyries
	2 Older greenstones
	1 Older granites

Bibliography

Beerstecher, Prof. E, jun., Petroleum-Mikrobiologie, New York 1954.

Berg, Prof. Dr. Georg, Vorkommen und Geochemie der mineralischen Rohstoffe, Leipzig 1929.

Beyschlag-Krusch-Vogt, Die Lagerstätten der nutzbaren Mineralien und Gesteine, Bd. 1, 2 und 3, Stuttgart 1914–1938.

Blum, Dr. J. Reinhard, Handbuch der Lithologie oder Gesteinslehre, Stuttgart 1860.

Blumer, Dr. E., Die Erdöllagerstätten, Stuttgart 1922.

Brinkmann, Roland, Abriss der Geologie, Bd. 1–2, Stuttgart 1950/54.

Clara, Max, Dr. med., Entwicklungsgeschichte des Menschen, Leipzig 1943.

Cloos, Walther, Die Erde ein Lebewesen, Stuttgart 1952.

— Kleine Edelsteinkunde, Freiburg 1956.

— Werdende Natur, Dornach 1966.

Drescher-Kaden, Die Feldspat-Quarz-Reaktionsgefüge der Granite und Gneise und ihre genetische Bedeutung, Berlin 1948.

Friedensburg, Prof. Dr. F., Die metallischen Rohstoffe, Stuttgart 1937–1954.

Gäa-Sophia, Jahrbuch der naturwissenschaftlichen Sektion der freien Hochschule für Geisteswissenschaft am Goetheanum, Dornach 1926, Bd. I. (Darin die Arbeit von E. Pfeiffer, 'Die geologische Erdenentstehung im Lichte der Geisteswissenschaft'.)

Grohmann, Dr. Gerbert, *The Plant,* Vol. 1., London 1975.

— *Die Pflanze,* Vol. 2., Stuttgart 1958.

Heide, Prof. Fritz, Kleine Meteoritenkunde, Berlin 1934 u. 1957.

Hofmann-Rüdorff, Anorganische Chemie, Braunschweig 1956.

Hoffmeister, Prof. Dr. C., Die Meteore, ihre kosmischen und irdischen Beziehungen, Leipzig 1937.

— 'Meteorströme' (Meteoric Currents), Leipzig 1948.

— 'Rückschau und Ausblick auf die Erforschung der Meteore' in: Die Naturwissenschaften, 3/1949.

Hoogewerff, H. Scheveningen, Grundlegende Arbeiten über die Entwicklung der Pflanzenwelt (Palaeobotanik) auf anthroposophischer Grundlage (unpublished).

Klockmann-Ramdohr, Lehrbuch der Mineralogie, Stuttgart 1936.

Knauer, Dr. Helmut, 'Das Antlitz der Erde', Goetheanum, Jg. 26, Nr. 20, 21, 23.

— 'Entstehung und Aufbau der Mineralien', ebenda Jg. 28, Nr. 48, 49.

— 'Die Entstehung der Erdrinde', ebenda Jg. 29, Nr. 10, 11.

— Erdenantlitz und Erdenstoffe, Dornach 1961.

Kolisko, Lilli, Sternenwirken in Erdenstoffen und Mitteilungen aus dem biologischen Institut am Goetheanum, Stuttgart und Dornach, 1927–1936.

The following are translated:—

— The Moon and the Growth of Plants. London, 1938.
— Spirit in Matter. Stroud, 1948.
— Capillary Dynamolysis. Brookthorp, Glos., 1943.
— Workings of the Stars in Earthly Substances. Stuttgart 1928.
— Workings of the Stars in Earthly Substance Solar Eclipse, 29.6.27. Stuttgart 1928.
— Workings of the Stars in Earthly Substances. Jupiter and Tin. Birmingham 1932.
— Gold and the Sun. Eclipse 20.5.47. Stroud 1947.

Lepsius, Prof. Dr. L., Biologische Betrachtungen im periodischen System der Elemente, in: Die Heilkunst, 65. Jg., Nr. 10.

Leunis-Senft, Synopsis der drei Naturreiche, Bd. Mineralogie und Geologie, Hanover 1875/76.

Lorenzen, J. Th., Metamorphosen, Hamburg 1958.
— Grundprobleme der Evolution, Hamburg 1958/60.

Pelikan, Wilhelm, The Secrets of Metals, New York 1973.

Pia, Julius, Pflanzen als Gesteinsbildner, Berlin 1926.

Polonovski, Michel, Medizinische Biochemie, Berlin-Saulgau 1951.

Rittmann, Dr. A., Vulkane und ihre Tätigkeit, Stuttgart 1936.

Schneider, Dr. Karl, Die vulkanischen Erscheinungen der Erde, Berlin 1911.

Stadnikoff, Prof. Dr. George, Die Entstehung von Kohle und Erdöl, Stuttgart 1930.

Stahmer, A. M., Erdölvorkommen der Welt, Mainz 1956.

Steiner, Dr. Rudolf, Arbeitervorträge 1922–24, Dornach 1959–1969 (8 Bde.).
— Mystery Knowledge and Mystery Centres. London, 1973.
— Occult Science, an Outline. London, 1962.
— Cosmic Memory. New York, 1961.
— Medical Lectures. 1920–24.
— Was hat die Geologie über die Weltentstehung zu sagen? und: Das Gesicht der Erde, Stuttgart 1949.
— Genesis. London, 1959.
— Agricultural Course, London, 1958
— An Occult Physiology. London, 1951.
— Das Verhältnis der verschniedenen naturwissenschaftlichen Gebiete zur Astronomie, Dornach 1926.
— Supersensible Man, London 1945.
— The Influence of spiritual Beings upon Man, New York 1961.
— Der Ursprung der Tierwelt im Lichte der Geisteswissenschaft, Basel 1948.
— Man as Symphony of the Creative Word. London 1970.
— The World of the Senses and the World of the Spirit, New York 1947.

(*The German titles are available for borrowing in English in typescript.*)

Stempell, Prof. Dr. W., Die unsichtbare Strahlung der Lebewesen, Jena 1932.

Substanzforschung, Beiträge zur, Bd. 1. Dornach 1952.

Suess, Eduard, Das Antlitz der Erde, Wien 1892, Bd. 1–3.

Vernadsky, W. J., Geochemie, Leipzig 1930.

Vreede, Dr. E., Anthroposophie und Astronomie, Freiburg/Br. 1954. (Available for borrowing in English as:—
Astronomical Letters, 1927–30.)

Wachsmuth, Dr. Günther, Erde und Mensch, ihre Bildekräfte, Rhythmen und Lebensprozesse, Konstanz 1952.
— Die Entwicklung der Erde, Kosmogonie und Erdgeschichte, ein organisches Werden, Dornach 1950.

Wagner, George, Einführung in die Erd–und Landschaftsgeschichte, Öhringen 1931.

Waldmeier, M., Ergebnisse und Probleme der Sonnenforschung, 1941.

Warburg, Prof. Dr. Otto, Schwermetalle als Wirkungsgruppen von Fermenten, Berlin 1946.

Weinschenk, Dr. Ernst, Spezielle Gesteinskunde, Freiburg 1905.

Wolff, Prof. Dr. Ferdinand von, Gesteinskunde, Die Eruptivegesteine, Pössneck 1951.

Chemiker-Zeitung, Neues aus der Chemie, Heidelberg 1955–1958.

Zeitschrift für Atomphysik, 22. Bd. 1. Heft, 1942. (Bengt Edlén u. B. Strömgren).

Index